Breath Taking

Stopping the Plunder of Our Planet's Air

EDNA ZEAVIN

R&E Publishers ❖ Saratoga, California

Copyright © 1992 by Edna Zeavin

All rights reserved.

No part of this book may be reproduced without written permission from the publisher or copyright holder, except for a reviewer who may quote brief passages in a review; nor may any part of this book be reproduced, stored in a retrieval system, or transmitted in any form or by any means electronic, mechanical, photocopying, recording or other, without written permission from the publisher or copyright holder.

This book is sold with the understanding that the subject matter covered herein is of a general nature and does not constitute legal or other professional advice for any specific individual or situation. Anyone planning to take action in any of the areas that this book describes should, of course, seek professional advice from lawyers and other advisers, as would be prudent and advisable under their given circumstances.

R & E Publishers
P.O. Box 2008, Saratoga, CA 95070
Tel: (408) 866-6303   Fax: (408) 866-0825

Book Design and Typesetting by Diane Parker

Cover by Kaye Quinn

Library of Congress Card Catalog Number: **92-54170**

ISBN 0-88247-925-3

CONTENTS

1. Future of Planet Earth 1
2. From Arrhenius to Schneider 11
3. Why is The Air So Dirty? 19
4. Global Warming Scientists 31
5. Missing Sky Scientists 51
6. Radiation Activists 69
7. Airborne Toxic Activists 91
8. Rain From the Sky 111
9. Effects of Forest Ecosystems on Air Systems 127
10. Environmental Groups—Audubon to Sierra 149
11. Governments and Politicians 163
12. Blueprint for Tomorrow 177
13. Bibliography 191
14. Environmental Organizations 203
15. Index 207

ACKNOWLEDGMENTS

I would like to acknowledge the help of my literary agent, Gerry B. Wallerstein for aid in putting the book into its final form.

This book relied heavily on scientific research, and I would like to thank the librarians at Penn State Environmental Resources Research Institute, the University of Arizona Science Library, and the Wilmot Library.

I worked extensively with scientists, environmental organizations, and environmental activists, and would like to thank them for their special help. Some organizations outstanding with their help were: Sierra Club, Lighthawk, Environmental Defense Fund and the Nevada Desert Experience. Individual scientists who shared of their time in editing portions of the manuscript were: Stephen Schneider of the NCAR, Center for Global Change University of Maryland scientists—Alan Miller, Stephen Leatherman, Alan Teramura; Syukuro Manabe of Princeton University, John W. Gofman, Ph.D., Professor Emeritus of the University of California, and Helen Ingram of the Morris Udall Center at the University of Arizona.

A number of individual activists helped: Dr. George Smith of Houston, associated with the Sierra Club, Jerry Tinianow of Columbus, Ohio of the Sierra Club, State Representative Maria Glen Holt of Maine, Herman L. Bogart of New Jersey, lawyers of the Environmental Defense Fund, Bruce Rich, Lori Udall, Stephen Schwartzman, and Helmut Ziehe of Clearwater, Florida.

Organizations who shared with me were the South Coast Air Quality Management District of Los Angeles, Lighthawk, Center for Global Change at the University of Maryland, Morris Udall Center for Public Policy, World Wildlife Fund, Earth Day 1990, United Nations, Nevada Desert Experience, and the Canadian Government.

Portions of the book are heavily edited; any mistakes in the remainder of the book lie with the author.

1 THE FUTURE OF PLANET EARTH

The breath of life, the air system that is life-essential for all of man and animal species, is becoming threatened through the misuse of air space. However, there is hope of meaningful restoration of many of our damaged air resources and this hope lies in individual citizens, environmental groups and governments who become aroused to the dangers of past inaction and lack of knowledge. An example of one citizen who cares is Harvey Clement, a small business owner in Orange County, California who is integrating a healthy business with healthy air. He owns Only Ovals, but unfortunately his manufacturing process emits smog-forming compounds, and because of these emissions he must comply with the Los Angeles Air Quality Management Rules. With a willingness to help the air quality efforts, he has made significant changes in his business.

Problems abound in air issues, but they can be arrested by concerned people. Global warming of the atmosphere can cause irreparable harm to agriculture, forests and oceans of the world. Holes in the ozone layer over the Antarctic signal potential future problems with agriculture, cancer for humans and marine life problems. Many animal and plant species are threatened with extinction by deforestation affecting the carbon dioxide and oxygen balance.

For years, acid rain has devastated the forests, lakes and streams of Europe, and now becomes equally destructive to the United States. Uncontrolled toxic waste production of manmade chemicals threatens the air that we breathe. Radiation threats continue with danger from Chernobyl-type nuclear reactors harnessed to give us electricity. And even our homes have been tightened for energy efficiency and seem to be a storehouse of toxic fumes—we no longer can safely retreat to our homes to escape the environment.

Positive actions by concerned environmentalists can mend some of this devastation! Governments are already swinging into action and have been aware for years of beneficial actions. In the early 1970's, the United Nations Environment Program (UNEP) was created. Three organizations, the UNEP, the World Meterological Organization and the International Council of Scientific Unions joined forces to study the greenhouse effect. Part of the greenhouse problem, nuclear fission was first regarded as an ideal energy solution until early in the 1970's when many doubts were raised as to the cost and safety of nuclear programs. Accidents at Three Mile Island in the U.S., at the Sellafield Nuclear Processing Center in the UK, and at Chernobyl in the Soviet Union undermined public confidence. We now believe that nuclear fuel may be potentially more dangerous than fossil fuel. Environmental groups are springing into action as well as governmental intercession.

Recently, the Sierra Club started a Global Warming Action campaign to address the frightening environmental crisis. They warn us that carbon dioxide, nitrous oxide, methane and chlorofluorocarbons ARE rapidly building in the atmosphere, and carbon dioxide and nitrous oxide are a direct result of fossil fuel combustion. Tropical rainforests are now being destroyed at a rapid rate and their effect as a sponge for excess carbon dioxide is removed from the environment. The Sierra Club recommends public education, safeguarding the world's rainforests, re-tooling the auto industry and forging a new national energy policy.

Many people remain unaware of indoor pollutants and their effects on our health and well being. Helmut Ziehe, recently moved from Germany to Florida, advocates the Baubiologie movement ("Bau" for building and "biologie" for life). This describes a movement to use woods and other natural materials in constructing and furnishing our homes. This will help the "sick building syndrome" affecting people's health and due to airtight, synthetically formulated buildings.

Helmut Ziehe is one of a growing army of involved citizens who are affecting our breath of life, and who are devoting themselves to cleaning up environmental air pollution. A number of their personal stories are included under the different chapters—some are working in conjunction with environmental groups. Some are concerned scientists involved in highly technical aspects like the stratospheric ozone layer, or radiation and others are concerned with politics and economic aspects of air pollution. There does seem to be a large group of involved lawyers, because they are used to dealing with litigations and political and governmental bureaucracies.

Our nation is trying to cope with the many pollution problems and has enacted a series of laws to curb pollution; the Clean Air Act, the Safe Drinking Water Act and the Clean Water Act. Some of the efforts are foundering due to lack of funds. The Superfund cleanup, first heralded as the solution to environmental woes, has become too cumbersome and results in temporary remedies. Only a few sites have been cleaned with thousands remaining as chemical time bombs. The Environmental Protection Agency lacks enough funds to do the proper job—in 1988 the EPA's workload was twice that of 1981, yet its purchasing power was the sam as 1975. Much dangerous contamination remains at nuclear weapons production facilities—it has been estimated that cleanup of these sites will cost more than one hundred billion dollars.

The next story illustrates the importance of everyone working together to solve toxic pollution problems and concerns three groups of people. Dr. James Ludwig, an environmental scientist works hand-in-hand with a concerned employee activist, Jane Elder. She in turn works in an environmental group office, The Great Lakes Sierra Club office.

The Great Lakes were once a pristine area and still contain one-half of all the fresh water on the earth. Thousands of years ago these lakes were shaped and created by the great ice sheet of the last glacial era. As the ice sheet advanced, it sheared the mountains and bared rock formations that originated at the beginning of geologic time.

Michigan, bordered by three of the Great Lakes, Superior, Michigan and Huron, was originally named by the Chippewa Indians who called this state Michigama meaning "great lake" or "large lake". The shores of Lake Superior lap on the upper peninsula boundary of Michigan and located in Lake Superior lies Isle Royale National Park. This is the largest Great Lake island and serves as a national park and game reserve. Conservationists on both the Canadian and U.S. sides of the border have struggled to keep this wilderness unspoiled. There are no autos allowed, no roads (but extensive hiking trails) and hunting of the last outpost of the North American timber wolf and moose are forbidden.

In the interior of Isle Royale lies an interior lake called Siskiwit Lake. Collections of water and fish in the summer of 1988 produced the shocking news that new toxic chemicals were being added to this once pristine lake. Dr. Deborah Swackhammer and Dr. Ronald Hites discovered manufacturing precursors of pentachlorophenol, mirex and octochlorostyrene along with the previous chemicals of PCBs and dioxin. These chemicals could only have come from distant manufacturing plants via atmospheric distribution.

During the same summer, Dr. James Ludwig found increasing rates of severe birth defects among cormorant chicks in the Upper Great Lakes including Lake Superior. Dr. Ludwig has studied colonial water birds for 20 years, and saw a sharp increase in deformities from the late 1960's to the early 1970's. In a three-year study of cormorants, he found 66 cormorants out of 10,335 chicks born with deformities. They exhibited such malformations as crossed beaks, hip deformities, clubfootedness, eye abnormalities leading to blindness, the failure of body walls to grow around the yolk sac causing body organs attached outside their bodies at birth.

But why should anyone be concerned over a few malformed birds? Because these problems are typical of many more issues concerning the ecological food chain; these directly concern people in the Great Lakes area. Rachel Carson in *Silent Spring* began her book with a fable of tomorrow, and told of a fictional place where the birds had either died or flown away, domestic animals were unable to reproduce or thrive, vegetation was brown and withered and lifeless streams had no more fish. She mentioned that not all disasters had occurred in one place, but some have happened to numerous other towns.

Our neighbor to the north, Canada, is concerned about the Great Lakes problem. Environmental Canada did a study over a wide array of species and discovered toxic pollution damaged the 16 top predator species with neurological effects and reproductive damage. The top predators in the Great Lakes aquatic food web exhibiting the highest contamination are the bald eagle, Caspian tern, herring gull and the lake trout. The effects concentrate on the viability of the embryonic newborn.

More proof is accumulating that toxic air pollution is a critical Great Lakes issue, and that one-half or more of the pollution in Lakes Michigan, Huron and Superior can be attributed to airborne distribution. The atmospheric pollution can potentially last up to five days and come from the Northeast, Midwest and industrial areas, Tennessee, and the Ohio River Valley, and the central breadbasket of the U.S. and Canada. Over 1,000 chemicals have been identified in the Great Lakes and include DDT, PCBs, dioxin, furans, aldrin, dieldrin and mercury in the food chain.

PCBs and DDT have both been banned in the U.S. and Canada, but still enter the lakes from worldwide sources. DDT is still used in the countries of Mexico, Central and South America and India, while the PCBs enter the lakes from existing sources such as landfills, spills, leaking transformers and contaminated sediments. Mercury comes from both natural sources in the rocks and soils and from coal burning facilities and incineration. The gaseous mercury in the atmosphere washes out in rain and snow in the presence of acids and oxidants, known as acid rain.

Mercury contamination is such a problem that the state of Michigan issued new fish advisories for all inland fish. "Pregnant women, nursing mothers, women who intend to have children and children age 15 and under should not eat fish contaminated at this level (0.5-1.5 ppm mercury) more than one meal a month." A similar study in 1984 followed pregnant women who ate PCB contaminated fish. They were divided into three groups on their level of fish consumption—low, medium, and high, and control groups. At birth the offspring of women were compared to the PCB exposed mothers; the children had a lower birth weight, were shorter, had a smaller head circumference and had behavioral and neuromuscular anomalies. Although the PCB children were the same as those whose mothers smoked or drank alcohol the PCB babies didn't catch up with their peers at a six month period.

What has been done to improve the toxic chemical air pollution in the Great Lakes area? A few steps have been taken, but more needs to be accomplished. The EPA has started a Toxic Air Monitoring System (TAMS) and it monitors the atmosphere at 11 stations across the country and in four cities—Boston, Chicago, Houston and Seattle.

The Great Lakes Atmospheric Deposition Network (GLAD), started by the EPA, identifies air pollutants that reach the lake—three stations are operating and all are located at Green Bay, Wisconsin. They hope to add three monitoring stations for each Great Lake. The existing stations monitor 20 metals and 8 nutrients.

Lastly, the University of Wisconsin Sea Grant program is funding the "Great Lakes Food, Fish and Human Reproductive Outcome" study in Green Bay, Wisconsin. They are studying the effects of consuming restricted fish by 804 women at a cost of $160,000 funded by both the federal government and the state of Wisconsin.

Traveling from Wisconsin and Michigan to the Western part of the U.S., there is an exciting story involving radioactive gases in Tucson, Arizona. This story illustrates how concerned citizens, employees, and government officials acted to eliminate a hazardous problem.

Radioactive pollution in the air is a common source of air contamination and is caused often by industrial manufacturing. The American Atomic Corporation plant in Tucson constantly emitted radioactive tritium (cancer causing gas) from its stacks. By June 15, 1979 the Arizona Atomic Energy Commission discovered high levels of this radioactive form of hydrogen in neighbor's lawns, fruit trees, swimming pools and urine. A normal level of tritium is 250 picocuries, but many of the neighbors had levels in their urine ranging from 34,600 picocuries per liter to 89,100 picocuries.

A company technician spilled tritium-laced oil over himself in 1978; it was discovered, he was fired and then complained to the state. At that time, the Arizona Atomic Energy Commission inspected the plant and warned of large tritium losses from the plant. Subsequently, it was discovered that the workers only wore rubber gloves and that the tritium alarm monitor went off every day causing workers to be evacuated from the factory. Estimated losses of tritium were 300,000 curies lost in 1978.

The American Atomics plant had been in business for 13 years, when in 1976 it started making tritium filled tubes for hand-held field compasses and gun sights for the military. Later, they extended their manufacturing to make tiny tritium filled tubes for back-lighting watches. Their employed workers rose to 200 in June, 1979 and their business rose from $500,000 in 1976 to $7 million in 1978.

One former worker, laser technician Elaine Hunter of Tucson commented, "We were inhaling tritium, so it was getting into our bodies." An estimated 300,000 curies of tritium were lost in 1978, the year of maximum losses and production efforts. Neighbors of the plant became very worried over their health with good reason. Dr. John Gofman, Professor Emeritus of Nuclear Physics at University of California in Berkeley remarks, "I estimate that cancer risk from radiation is 35 times as great as the official estimates of the nuclear community. Tritium is a very low level emitter, so it is even more harmful—1-1/2 to 5 times more harmful than any other beta emitter. Tritium emits beta radiation which is as damaging as gamma radiation or X-rays inside the body."

At this point, an anti-nuclear group, Nuclear-Free State conducted a protest outside the plant gates with 150 people protesting. On June 2, 1979, the Arizona Atomic Energy Commission scheduled Tucson hearings. Between June 11th and 15th, state politicians, Tucson City Council, Pima County Health Department Director and Governor Bruce Babbitt called for the plant's closure. On July 11th, American Atomic officially surrendered their license.

Later, the Tucson School District was surprised to find $300,000 worth of fresh food and $270,000 of canned food contaminated with tritium. The school kitchen closed June 1st for more than a year and students were forced to eat sack lunches. The nearby Senior Now Generation, who prepared lunches for 1,400 elderly people, had to close their kitchens after tests found 40,000 picocuries of tritium per liter of water in gelatin.

TUSD was forced to bury the tainted food in a deserted World War II bombing range 25 miles southeast of the city. Governor Bruce Babbitt sent the Arizona National Guard in September, 1979 to seize 550,000 curies of tritium stored in 55 gallon drums. The tritium then traveled to Navajo Army Depot in Flagstaff, to Mounds Lab in Ohio and subsequently to Oak Ridge National Lab in Tennessee.

What happened to the original tritium study commissioned by the federal government to track the health of plant employees? Sad to say, it was lost in the federal bureaucracy and the tissue slides completely disappeared. The study of 25 most-exposed American Atomics workers went to a agency called National Institute for Occupational Safety and Health, and was never recovered.

In spite of mistakes and errors at all levels of government, we can be guardedly optimistic about environmental issues. The developing countries are becoming more aware of the importance of the environment to their economic development. We must change the pattern of energy use and develop a new concept called energy intensity of economics. This is defined as how much energy you have to produce a dollar of GNP. Many developing countries are using new agricultural methods called agroforestry where tree and crops grow together. The Chinese have made awesome progress by planting trees by the roads.

2 FROM ARRHENIUS TO SCHNEIDER

Disastrous air pollution has continued in London for centuries, since the reign of the first Elizabeth when coal began to be the fuel of choice. Some of the worst events happened when there was a prolonged fog with temperature inversion preventing the dispersion of sulfur dioxide. In 1872 a portent of future events took place when a prolonged smog killed several thousand people.

The London "Black Fog" of December 5-9, 1952 caused 4000 deaths in four days making the city completely nonfunctional. This event remains the most serious air-pollution in history, even worse than the Second Battle of Ypres in World War I where chlorine gas clouded the atmosphere. It started on a Thursday, December 4th as a white fog like the characteristic pea soup of London, but rather more dense. Buses ran at two miles per hour with the conductor walking in front to guide the driver. Even the animals had problems at the greyhound track when the dogs lost sight of the rabbit.

The fog affected all kinds of transportation, and one pilot got lost taking passengers to the airport terminal after landing safely. Most of the populace took the subways for their transportation needs. By Friday, the color of the fog changed from white to brown, and when Saturday morning arrived the color became black. At this point in the catastrophe, people started to die rapidly. At noon on Saturday, all the hospitals were full and by Sunday morning the visibility was so poor that people couldn't see their hands in front of their faces. Many old people became lost in the smog and died of exposure from acute respiratory attacks.

Learning from this sad experience, the British health authorities enacted the Clean Air Act of 1958. This Act eliminated much of the smoke caused by burning soft coal, and alleviated the bad effects of sulfur dioxide, smoke and fog.

Down through the centuries, air pollution as illustrated by the "Black Fog" remains the norm for man with the progress of civilization. We're now reaping the consequences of the Industrial Revolution and our extreme reliance on fossil fuels for transportation and industry.

However, hope is arising from an array of different governments, conservation groups, scientists and individuals as they tackle our air environmental problems. A comprehensive history of the important events will allow us to learn from past history and progress to the decade of the 1990's and the twenty-first century. The average citizen can learn to make small contributions of his own to allay the effects of environmental air consequences.

Air pollution problems have been building for thousands of years, since the philosopher Seneca noted pollution in Rome in A.D. 61. Later, mention was made of an air problem from the burning of coal in England by Eleanor of Aquitaine, the wife of King Henry II about one thousand years after Seneca's comments. The coal-engendered pollution became so bad that in 1273 a royal decree forbade the combustion of coal in London.

However, the most pressing air problem began with the Industrial Revolution in the eighteenth and nineteenth centuries in Western and Central Europe, England and the United States. The advent of steam power for commercial purposes decreased the air quality. Since London suffered from extensive air pollution, by 1819 the British Parliament undertook studies of pollution abatement. Later in the 19th century, Tyndall prophesied of the future greenhouse effect by stating that carbon dioxide in the atmosphere together with water vapor absorbed longwave radiation.

The father of the modern air movement was Svante August Arrhenius, a Swedish physicist and chemist, who won a Nobel prize for his work in electrochemistry and became director of the Nobel Institute for Physical Chemistry in Stockholm, Sweden. In 1896, he wrote an article for the "Philosophical" magazine and calculated that a doubling of atmospheric carbon dioxide would raise global temperatures by 4° C. He believed that the mass of carbon dioxide in the atmosphere is only a fraction of that which existed in the past and much was absorbed from the atmosphere by the formation of carbon containing rocks. He felt the carbon dioxide concentration of the atmosphere could vary over a wide range.

Around the same time in 1899, P.C. Chamberlin worked on the carbon dioxide balance in the atmosphere. He believed the amount of carbon dioxide coming from the earth varied according to the level of volcanic activity and other natural factors. In the past, with glaciation and the elevation of continents, this led to an increase in the erosion of surface areas in weathered rock and decreased the atmospheric carbon dioxide concentration.

By the 20th century in the United States, the Industrial Revolution was in full swing. New sources of pollution included factories mushrooming all over the land, electrical power generating stations and the advent of motorized railroad and auto transportation. Besides the United States involvement in the study of air problems, the Russians were equally involved. The global environmental change problem had its roots in the 1920's when the Russian mineralogist, Vladimir I. Vernadsky wrote on the biosphere. In 1925, A.J. Lotka associated the industrial use of fossil fuels with increasing carbon dioxide in the atmosphere.

The first warning of future effects of climate change came with the predictions of G.S. Callender in 1938. He proposed that an increase in atmospheric carbon dioxide concentration due to man's economic activities would result in global warming. Attempting to estimate climatic change caused by man-made carbon dioxide to the year 2200, he felt the burning of fossil fuels would lead to a tenfold increase in atmospheric carbon dioxide.

With the advent of the scientific effort on global climatic change, finally the U.S. government started to address this problem. They passed an Air Pollution Control Act of 1955 aiming to provide research, technical and financial assistance to states working on pollution problems. A warning was sounded by R. Revelle and H.E. Suess of the Scripps Institute of Oceanography in 1957. They said, "humanity is performing a great geophysical experiment, not in the laboratory or computer, but on our planet." They predicted most carbon dioxide from burning fossil fuels would remain in the atmosphere and not be absorbed by oceans. It fell to Charles D. Keeling of the Scripps Institution of Oceanography in La Jolla, California to back the global warming effect with solid scientific data. He began monitoring carbon dioxide levels at an observatory 11,000 feet on top of a bleak, extinct volcano of Mauna Loa in Hawaii. He selected this site so that the findings would not be swayed by carbon dioxide uptake and release of oxygen by green plants during photosynthesis and respiration. The scientific records covered 24 hours a day for every day for 30 years. The data specified that the levels of carbon dioxide had gone from 315 parts per million, 30 years ago to 348 parts per million, today, or an increase of 0.4 percent a year.

The decade of the 1960's began with more U.S. Government intervention in air pollution problems with the Clean Air Act of 1963. This Act Expanded the federal government's role in intermunicipal and interstate air quality control. It tried to provide federal funding, but the focus dealt only with motor vehicle emissions. The next large federal legislation was the Air Quality Act of 1967. With this Act, the federal government expanded its role, but control of air pollution was still felt to be a state and local government responsibility. The best qualities of the act authorized the federal government to designate air quality control regions, recommended specific control technologies and conducted expanded research and development. Besides air quality control, by late 1960's, there was an advent of computer modeling of the greenhouse effect.

The federal government started playing an ever-expanding role in air quality when it passed the Clean Air Act Amendment of 1970. At the same time the Environmental Protection Agency was formed and consolidated many scattered antipollution programs. The federal government achieved additional power to deal with air quality problems at this time. These amendments let the federal government set uniform ambient air quality standards for some pollutants, established uniform emission standards for new sources of pollution, and required the states to formulate state implementation plans to attain standards. During the same year of 1970, more warnings surfaced about the potential greenhouse warming of the earth by the scientist, Manabe. He felt that by the year 2000, the mean air temperature would be 0.8° C higher than in 1900 due to the carbon dioxide growth in the air.

The first global consciousness of the environment occurred in the U.N. Conference on the Human Environment held in Stockholm Sweden in 1972. The U.N. Environment Program (UNEP) was developed to help promote sustainable development of the environment. Closely following the development of the UNEP came the first research on the distribution of CFCs (chlorofluorocarbons) around the world. Jim Lovelock, a British scientist, investigated the distribution of the CFCs and published his results of the measurement of the concentration of CFCs in the air in the magazine, "Nature" in 1973.

Immediately after Jim Lovelock's research came the findings about CFCs by a team of scientists from the University of California at Irvine, Sherry Rowland and Mario Molina, Ph.D. They discovered that CFCs do not interact with living things, don't dissolve in the ocean and don't get washed out of the air by rain. It's hard to imagine, but they float around until they work their way into the stratosphere. Here, the CFCs are broken apart by ultraviolet radiation and chlorine atoms are released. These chlorine atoms can scavenge and destroy many thousands of ozone molecules and destroy the shield protecting the earth from ultraviolet rays. These scientists said that if CFCs are released at the rate of 800,000 tons a year (rate of release in 1972) within 30 years, half a million tons of chlorine in the stratosphere would destroy 20-40% of the ozone shield.

In 1976, the National Academy of Science reported on CFCs using the scientist V. Ramanathan's work. If the release of CFCs grew at 10% a year, their influence could exceed that of carbon dioxide produced by human activity by the end of the twentieth century. By increasing the amount of CFCs to the atmosphere at a growth of even a few percent a year, it could lead to a climate change of drastic proportions.

In 1979, 32 European nations, Canada and the U.S. signed the Convention on Long Range Transboundary Air Pollution pledging to reduce 1980 sulfur emissions by 30% by 1993. A number of nations pledged a 40-50% reduction in sulfur emissions and these included Canada, West Germany, France, Norway, Sweden and Denmark.

Continuing the clean air momentum of the 1970's, a new decade arrived signalling even more clean air involvement. In the U.S., we passed the Acid Precipitation Act of 1980 which created an interagency task force on acid precipitation and authorized a ten-year national acid precipitation assessment program. In 1980, the U.S. government established the Superfund to clean up toxic dumps.

By October, 1985, the first scientific consensus on global warming was achieved in Villach, Austria. This meeting involved several scientific organizations— the World Meteorological Organization (WMO), the U.N. Environmental Program (UNEP) and the International Council of Scientific Unions (ICSU). Scientists from 29 industrialized and developing countries concluded that climate change must be considered a possibility.

The same year, D. Raynaud used ice core analysis to relate the increase in atmospheric carbon dioxide to the burning of fossil fuels. Previously in 1982, strange events were unfolding in the Antarctica when a team of researchers from the British Antarctic Survey working at Halley Bay in Antarctica found an odd depletion in ozone in the air above the Antarctic. J.C. Farman in an article in the "New Scientist" in 1987 said that the depletion of ozone is more than 50% over a period between 30-40 days a year. He called for a reduction in CFC emissions by 85% as quickly as possible. He advised an immediate reduction of CFCs for all but the most essential applications, such as refrigeration.

Concurrently, the Dahlem Workshop took place in Berlin in November, 1987 and involved a number of ozone scientists, Lovelock, Rowland, Crutzen and the Antarctic scientists Farman, Philip, Solomon and Watson. At this meeting, it was established that chlorine from industrial products, especially CFCs are to blame for the destruction of the Antarctic spring. The main hole in the ozone layer centered on the South Pole with smaller "miniholes" appearing a few days and then disappearing. During the long polar night, chemical processes occurring naturally nowhere else on the earth take chlorine atoms out of reservoirs and release them. When sunlight returns in spring, the photochemical processes begin to work causing the chlorine's release.

Montreal, Canada on September 16, 1987 became the site of the first global treaty called the Montreal Protocol aimed at reducing air pollution. 46 nations pledged to freeze CFC production at 1986 levels in 1990 and cut CFC production 50% by 1998. They agreed to freeze consumption of halons by 1992—halons are three to ten times more destructive than CFCs. James Hansen, on June 23, 1988 (head of NASA's Goddard Institute of Space Studies) predicted that the evidence is strong that the greenhouse effect is here. "It's time to stop waffling so much. The evidence is pretty strong that the greenhouse effect is here." Steve Schneider, climatologist at the National Center Atmospheric Research in

Boulder, Colorado remarks, "If our projections are correct, during the next century Earth may warm 10 to 40 times faster than it did after the last Ice Age."

There were a few loopholes in the Montreal Protocol and two years later at a conference in Helsinki, Finland, 80 nations gathered to strengthen the 1987 Montreal Protocol on chemicals that damage the ozone layer. The nations will phase out production and consumption of chlorofluorocarbons as soon as possible and no later than the year 2000. The same year, 70 nations met in the Netherlands in November 1989 to try to stabilize carbon dioxide emissions.

By February, 1990, the key senators and Bush administration officials reached an agreement to clean up foul air. These changes would be the first adopted by Congress since the Clean Air Act of 1977. Currently, the Senate bill would reduce dangerous smog in cities by imposing tighter emission standards and cutting industrial and residential pollution. The bill would set standards for 200 unregulated toxic air pollutants and cut sulfur dioxide emissions from power plants to combat acid rain. Of great importance, it would phase out ozone depleting CFCs by 2000 and reduce carbon dioxide emissions.

With a sound background of the history of air pollutants, the next chapter will begin to address the various air pollution problems that are facing the planet. Hopefully, the reader will start to see ways in which he can help air pollution, not only in small ways but in larger areas. Everyone can get involved in these issues. All through this book, there will be many ideas advanced as governments, environmental groups, and individual citizens act to achieve our clean air goal.

3 WHY IS THE AIR SO DIRTY?

In the petrochemical corridor of Texas (five Texas Gulf Coast Counties), the chemical plants and oil refineries are an everyday risk. An elementary school teacher remarked, "Every time I drive down (highway) 225 and see all the plants and big storage tanks, it always occurs to me, if it would blow up, I would die on the freeway." An accident did happen in October, 1989 when the residents of Pasadena, Texas were shaken by a blast at the Phillips Petroleum Company plastics plant. This explosion caused at least one death and resulted in 23 missing people.

Over the past seventy years, many other industrial accidents have happened around the world. The MIC gas leak in Bhopal, India illustrates the worst industrial accident in history. Seveso, Italy experienced an uncontrollable exothermic reaction in a reactor at the Hoffman-La Roche Givaudan chemical plant causing an explosive release of dioxin. This accident contaminated 4,450 acres of land and killed over 100,000 grazing animals. In 1986, a fire at a chemical warehouse in Basel, Switzerland released toxic pesticides into the Rhine River.

Industrial accidents are just a few of the environmental air problems encountered by the average citizen. In order to become fully aware of the steps being taken by governments, environmental groups and individuals, we must define and look at various environmental air problems both here and throughout the world. A very showy and prominently displayed air problem , the greenhouse effect, is being worked on by scientists and governments all over the world. Less showy, but still present in ever increasing amounts remains the toxic air problem associated with mobile and non-mobile burning of fossil fuels. Acid rain, an issue with tremendous environmental effects, is affecting wildlife, especially fish, trees and other plant life and has a deleterious effect on human health. It destroys and is destroying magnificent historical monuments and buildings.

Industrial chemicals in the air are a product of our society and the continuing greed for more money. Finally, Chernobyl made us aware of the problem of both massive and low level radiation from nuclear power plants, and mining accidents of radioactive ore disposal, and disposal of radioactive waste material.

Internationally, alarm is arising about the greenhouse effect and the blame rests squarely on the combustion of fossil fuels (coal, oil and natural gas). The rise in carbon dioxide in the air, the main greenhouse gas, seems directly due to the ever increasing use of fossil fuels. In fact, fossil fuels are responsible for 70% of all human carbon dioxide emissions, or 5.6 billion metric tons of carbon each year.

Other gases, also, contribute to global warming: nitrous oxide, methane and indirectly tropospheric ozone (component of smog.) How do these gases occur from fossil fuels or their burning? Nitrous oxide results from combustion, methane originates from leaks in natural gas wells, pipelines and coal mines, and ozone arises from chemical reactions involving methane and is an element of photochemical smog.

Since the Industrial Revolution in the 1800's, carbon dioxide is being released in larger quantities than plants and oceans can absorb the gas. Increases of carbon dioxide accelerated rapidly in the early 1900's when people began to burn gas and oil. The rapid destruction of tropical rain forests has caused a great release of carbon dioxide through either burning or the rotting of logs from the forests.

Other troublesome greenhouse gases include CFCs (chlorofluorocarbons) found in refrigerants, aerosol sprays, foam products and solvents; nitrous oxide generated from nitrogen rich fertilizers in agriculture and from the combustion of fossil fuels, and methane arising not only from natural gas, but coming from many living sources as well. Methane originates from burning biomass, emissions from solid waste dumps, belching of cows, and waste gas from termite activities.

Scientists from many countries continue to hold conferences to warn us of the potential problems of global warming. Michael McElroy, chairman of the Department of Earth and Planetary Sciences at Harvard University addressed a group of international scientists at a conference in Toronto, Canada. He remarked about carbon dioxide, "It is the single, largest waste product of modern society. Approximately half the carbon since the Industrial Revolution persists in the atmosphere, today."

Charles Keeling of the Scripps Institute of Oceanography measured the annual increase in atmospheric carbon dioxide since 1958. To slow global warming would require a 20% cut in worldwide fossil fuel consumption—to stop it would require a 50% reduction back to the levels of the early 1960's.

WHAT IS THE GREENHOUSE EFFECT?

The rest of the world is part of an even larger scientific experiment—what will happen to our planet when we change our atmosphere? Over the last hundred years, the average global temperature rose nearly one degree Fahrenheit. The average citizen experiences the greenhouse effect when he opens his car doors on a hot day. The car windows allow sunlight in and trap the heat inside the car just like the greenhouse gases, for example carbon dioxide traps the sun's heat within the Earth's atmosphere. The greenhouse effect makes our planet livable.

Man causes many problems and he generates waste gases of industrial activity upsetting the equilibrium of our greenhouse. Through fossil fuel consumption and deforestation, we have increased the atmospheric carbon dioxide concentration 30% in the last 30 years. As the greenhouse gases continue to buildup, more heat becomes trapped in the Earth's atmosphere.

The largest part of the greenhouse effect is caused by carbon dioxide which accounts for 50% of the greenhouse effect. New gases in the greenhouse scenario, the chlorofluorocarbons, cause large openings in the ozone layer protecting the earth from ultraviolet radiation. Additionally, they contribute to the greenhouse effect by being thousands of times more heat absorbing. Several world conferences were held to ban aerosol cans chlorofluorocarbon propellants, but other activities still contribute like refrigerants, air coolers and styrofoam packaging.

FOSSIL FUEL USAGE

Since the consumption of fossil fuels is responsible for the greatest share of human emissions of carbon dioxide, the best and most cost-effective and environmentally sound approaches are to become more energy efficient through using alternative fuels and to use renewable energy sources. Charles L.Gray, director of the EPA's emission control technology division in Ann Arbor, Michigan said, "Although we have made significant progress in reducing emissions per mile traveled over the last 20 years, the number of cars and the miles they travel has almost doubled in the same time. Unless we continue to reduce the pollution emitted per car, clean air in many cities will continue to be an illusion of the dreams of the '60s and '70s."

An energy-efficient way to tide the emissions of carbon dioxide and carbon monoxides may be through alternative fuels such as methanol, gasohol and other oxygenated fuels. Phoenix, Arizona, long known for its clear skies, has become increasingly polluted with emissions from cars of over a million residents. In the fall of 1989, Phoenix tried an oxygenated-fuels program to reduce carbon monoxide poisoning which is at its worst during the winter. The gasoline is oxygenated by blending it with an oxygen-enriched additive, an ether called methyltertiary-butyl ether—MTBE, or with ethanol (an alcohol.)

The Department of Environmental Quality faced the monumental task of dispensing to the public the information about the new fuel, and were allocated $500,000 for public education. Soon, a TV and radio jingle appeared on the airwaves, "We've got a solution for pollution for drivers who care, solution to pollution." Other ways they advertised besides the media were through billboards and consumer pamphlets and through hotline numbers.

What effect did the oxygenated gas, MTBE have on the maintenance and running of cars? The effect varied by model, make and year with cars built since 1982 able to handle the new fuels. The fuels can wreak havoc with fuel systems in older or poorly maintained cars, and make the car hesitate or stall. Another problem is with stalling or hesitation in hot weather.

Besides increasing energy efficiency in fossil fuels, the Union of Concerned Scientists believe renewable energy sources will solve problems in the future. There are inexhaustible resources of sunlight, winds, oceans, rivers and plants and with technological advances these sources promise to replace fossil fuels. Renewable energy sources have many advantages over fossil fuels and nuclear power. They produce little or no pollution or hazardous waste and generate few risks to public safety. Renewable energy can be domestically generated in the countries throughout the world making it immune to foreign disruption like oil embargos. Renewable energy sources are enormous, for example 44,000 quadrillion British thermal units (BTU) or quads of sunlight fall on the United States each year which is equivalent to 500 years of current U.S. energy consumption. Winds represent 3,000 quads per year of kinetic energy, 100 quads per year in tree and plant growth and six quads per year in rivers. Geothermal power, or heat in the earth's crust can amount to one million quads, or over 15,000 years of current energy use.

SAVING ENERGY COSTS

Energy use from cities with large populations contributes to global heat. Over the last 200 years, the world population has increased from one to five billion people, while future populations will stabilize in the next century to between 8-14 billion people. In 1950, Paris and New York were the only cities with populations of more than 10 million, while in 1975 there were seven cities of this size and more than half of the world's population lived an urban life. Cities contribute to the greenhouse effect by using fossil fuel for transportation, electricity and industry.

"Time" magazine in their January 2, 1989 issue featured a cameo illustrating how one community became more energy efficient. Osage, Iowa, a town of 3,600 people started to save energy the old-fashioned way by simple measures. Some of their measures include plugging leaky windows, insulating walls and ceilings, replacing inefficient furnaces and insulating hot water heaters. The Osage Municipal Utilities recently gave customers #15 fluorescent light bulbs which use much less energy than incandescent lights. Who is responsible for all these energy-saving measures? Weston Birdsall, the general manager of Osage Municipal Utilities, is the person behind the conservation measures. Since 1974, the community has cut natural gas consumption 45% and reduced annual electricity demand by more than half.

The "Audubon Activist" of January/February 1990 gave a worksheet to calculate your greenhouse gas activities, or the amount of carbon dioxide and CFCs emitted by your everyday activities. Some of the categories included numbers of motorized vehicles driven with miles per gallon and number of miles per year driven, household items used such as garbage and paper products, home heating and cooling including water heating, lighting and appliance use. They estimate if each of us cut our greenhouse emissions by 2% a year (or 1100 lb) by the year 2,000 we could achieve a 20% reduction of gas emissions.

ACID RAIN

Several of the by-products of fossil fuel consumption are sulfur oxides and various oxides of nitrogen, specifically nitrogen dioxide (NO_2) and nitric oxide (NO) labelled NO_x. When fossil fuels are burned, the sulfur and nitrogen in the fuel combine with the oxygen in the atmosphere to form sulfur and nitrogen oxide. These oxides fall as fog, sleet, snow or rain and may be either wet or dry.

The Clean Air Act has worked steadily since 1970 to reduce pollution from new cars and power plants, but loopholes in the law have allowed pollution from old power plants and smelters to be ignored. Many utilities have built over 217 tall smoke stacks that take pollution away from a local area and dump it in the upper atmosphere. It may fall as acid rain hundreds of miles away from the original source and threaten our health, natural resources and erode buildings, bridges and monuments.

Acid rain statistics continue to threaten us and have made great inroads on our aquatic resources. Six areas in the U.S. should be concerned—New England, Southeast Adirondacks, the Atlantic coastal plain, the Upper Peninsula of Michigan, the Appalachians and northern Florida. Acid rain kills fish directly by acidifying lakes and destroys other wildlife by killing the aquatic food chain. The 212 Adirondack lakes have had their fish killed by acid rain making the lakes too acidic for aquatic life. The Pennsylvania Fish Commission estimates that 6,500 miles of prime trout streams can be lost, and will result in an economic loss of $125 million in recreational income.

The effects of acid rain on forests are predictable. The rain strips the waxy coating from leaves, leaving the leaves vulnerable to sunburn, ozone burning, insects and fungi. The acidic snowpack can dissolve toxic metals from rock and soil to poison the trees at their roots. Many maple trees have been lost in Canada, and Germany has found over half of their nation's trees affected by Waldsterben (forest death) with a loss of $250 million annually.

Acid rain can threaten human health and is the third largest cause of lung disease after active smoking and passive smoking. The Office of Technology Assessment (OTA) determined that airborne sulfate particulate pollution is responsible for 50,000 premature deaths annually. Canadian scientists Dr. David Bates and Dr. Bonnie Stern work actively on the effects of acid rain on human health. Dr. Bates was present in the London killer fog of 1952, and since 1977 has been a professor at the University of British Columbia. He found a significant relationship between hospital admissions during the summer and the levels of sulfates and ozone in the air.

The United States' symbol of liberty, the Statue of Liberty, became badly eroded by acid rain. When the Park Service began to repair the statue, they found that acid-rain, other air pollutants and salt air had wreaked havoc on the steel and copper monument. Famous buildings throughout the world are literally melting away from the onslaught of acid rain, and include the Taj Mahal in India, the Acropolis in Greece, the Tower of London, and the Cologne Cathedral in Germany.

So far, this chapter has dealt with fossil fuel products, and combustion products of fossil fuels. Some air toxics are not combustion products, but have contributed to polluted air problems. Some organic molecules come from chemistry labs, for example chlorine-containing pesticides (chlorinated hydrocarbons) such as DDT and the herbicide 2-4 D, some related compounds namely PCBs (used in electrical transformers) and two other chemical groups of insecticides, organophosphates and carbamates.

Several devastating problems resulted with exposure to these chemicals, giving an acute toxicity as demonstrated in the 1984 Bhopal disaster caused by Union Carbide of India and involving the pesticide, Sevin. Others of these compounds persist in the environment, for example chlorinated hydrocarbons and do not break down by biodegradation. DDT, a common pesticide, has an affinity for fat and is taken up into living organisms and retained by crops and animals grown for human food. The U.S. took steps to ban DDT, but the compound is still used by Third World Nations, who in turn export their food products to us.

Biological food chains exhibit the worst effect of chlorinated hydrocarbons. The compounds accumulate and build to high levels in some wildlife, a process called bioconcentration. They concentrate into the fat-containing part of living things, and move upward from one trophic level to the next, that is plants to herbivores to carnivores and higher in the food chain. These compounds caused reproductive failure in birds, killed fish and stopped photosynthesis in marine algae. Our national bird, the eagle, was adversely affected as well as hawks, pelicans and fish.

An example of an acute toxicity was caused in West Virginia in August, 1985 and was precipitated by the Union Carbide Co. A large plume of toxic gas wafted over Institute, West Virginia but people weren't alerted soon enough. Imagine tens of people weeping and choking simultaneously. The alarm finally sounded, but by then 135 people required hospital treatment. The EPA later disclosed that 28 minor leaks of the deadly gas, methyl isocyanate (MIC) had occurred over a period of five years. Union Carbide actively engaged in manufacturing MIC at Institute for seventeen years, and shipped this chemical product to plants that made MIC based pesticides, Sevin and Furadan.

Besides chlorinated hydrocarbons and insecticides, heavy metals may be emitted into the air from leaded gasoline and from industrial factories. Lead from leaded gasoline can be changed to an inorganic form from the auto's internal combustion engine. The inorganic form of lead is a cumulative poison if inhaled or ingested and can be absorbed or stored in the body mainly in bones. Barry Commoner, a biologist and organizer of the first Earth Day in 1970, said there has been a 92% decline in airborne concentrations of lead between 1975 and 1987 after leaded gasoline was banned. Other heavy metals are emitted into the air from industrial sources. Cadmium comes from smelters, nickel can cause cancer and dermatitis, mercury could be quite dangerous and the vapor can cause neurological damage. Beryllium and molybdenum are toxic metals associated with metal alloy hardening and smelting and are responsible for farm animals deaths.

The last group of air pollution products includes low level ionizing radiation, such as those products released at the Chernobyl nuclear plant accident and radioactive isotopes which are an unstable form of elements undergoing radioactive decay. Ionizing radiation can break apart the delicate biochemical bonds in living tissue leading to chemical and biological mayhem according to Dr. John Gofman, Professor Emeritus of Medical Physics at the University of California in Berkeley. He comments, "With ionizing radiation, electrons are removed from their atoms and endowed with huge energy—they can break anything, anywhere. Once it has ripped an electron out of an atom in a molecule, the molecule is at a high energy level and can produce chemical reactions."

Ionization may not be a familiar term to most people, but it refers to the process of removing electrons from an atom. One of the simplest chemicals, hydrogen, has one electron carrying a negative charge in its orbiting ring, and the nucleus has a positive charge. When the electron is removed from the outer ring, the hydrogen atom will become a positive hydrogen ion. Radioisotopes were involved in the Three Mile Island accident in 1979 which gave off xenon, and this gas illustrates radioactive decay, since xenon has a half-life of nine hours. (One half of the radiation was released in that period).

One of the insidious effects of Chernobyl was the problem of contaminated soil which later affected the food supplies and contaminated them. Three years after the accident, 250,000 people live on soil so contaminated that they can't grow food. The radioactive cloud drifted over the European continent and subsequently many crops had to be destroyed. Additionally, drinking water was affected and poisoned streams produced contaminated fish.

Iodine-131 contaminated the milk in many areas of the Soviet Union causing the Soviets to use agrotechnical and agrochemical measures to make the food products fit for consumption. They had to resort to deep ploughing and use large quantities of inorganic fertilizers.

Dr. John Gofman warns that all forms of human cancer can be induced by radiation, secondly an increase in radiation produces a linear incidence in cancer and thirdly, children are most susceptible to radiation-induced cancer. His research showed the permissible annual dose of radiation from nuclear power plants—170 millirems per person could result in 16,000 to 32,000 added deaths from cancer in the U.S.

4 GLOBAL WARMING SCIENTISTS

James Lovelock, chemist-biologist-inventor, is so convinced of the importance of the greenhouse effect that he gave up university life many years ago to reside in his English countryside home. He says, "As a university scientist, I would have found it nearly impossible to do full time research on the Earth as a living planet." He formulated the Gaia theory which hypothesizes that living things have taken control of the Earth, and have transformed an inert chemical ball into an immense self-sustaining organism. The word Gaia for this theory comes from the Greek Earth Goddess. Lovelock warns that mankind is now doing everything to raise the carbon dioxide concentration by burning fossil fuels, producing gaseous by-products and destroying rich tropical vegetation that removes carbon dioxide from the air.

His research indicates tiny marine organisms called plankton help to regulate global temperatures. Some species of plankton produce a chemical, dimethyl sulfide (DMS) that accumulates in the ocean and permeates the atmosphere. The DMS oxidizes and leaves sulfate particles serving as condensation nuclei forming the basis for cloud formation. The amount of DMS produced by plankton may affect the earth's temperature by controlling the density.

James Lovelock isn't the only scientist who believes in the greenhouse effect! It has been corroborated by Syokuro Manabe of the Geophysical Fluid Dynamics Laboratory of NOAA located at Princeton University. Manabe is one of the original computer modeling scientists of the early 1960's, who predicted the warming of the earth by fastidious use of computers. Currently, he works with a CRAY YMP supercomputer for predicting models of future atmospheric climates. By doubling the present concentration of carbon dioxide, he generates temperature and soil maps of the world. In fact, this projected map of soil moisture resembles the maps of the United States during the 1930s with similar patterns. He states, "What I am saying is that the kind of summer drought we get from our model calculations may be similar to the drought of the Dust Bowl period."

Manabes's experiments indicate that an increase in atmospheric carbon dioxide not only alters the temperature, but affects the global weather cycle. His research predicts possible potential effects in the weather as follows. There will be an increased rate in the global mean rate of both precipitation and evaporation, a larger increase in runoff in crater river basins, an earlier arrival and ending of the snowmelt season, more frequent mid-continental summer dryness in the middle and high latitudes, increased soil moisture in middle and higher latitudes during the winter and a winter reduction of soil moisture in the poleward periphery of subtropical desert.

Overwhelmingly, scientists are in agreement with Lovelock and Manabe, although a few scientists offer a dissenting voice and disagree with the possibility of future greenhouse effects. The scientific method usually involves years of work, and most scientists don't publicize their work. However, the greenhouse phenomenon is defining the role of scientists through the new role of advocacy scientists. Helen M. Ingram, Director of the Udall Center for Studies in Public Policy, and Carole Mintzer have written a paper, "How Atmospheric Research Changed the Global Climate." Research center scientists are becoming very vocal in publishing the results of their research finding and two of the most prominent are Stephen Schneider, head of the Climate Systems at the National Center for Atmospheric Research in Boulder, Colorado, and James Hansen of NASA's Goddard Institute for Space Studies in New York City.

How do these scientists go about publishing their important work on the greenhouse effect? Schneider wrote a book on the greenhouse effect published by Sierra Books, *Global Warming: Are We Entering the Greenhouse Century?*, and testifies regularly before Congress. Additionally, he travels on the lecture circuit that includes foreign capitols.

James Hansen follows the same route as Schneider with testimony given before Congress, and an ability to give very apt quotes about the current climate. On June 23, 1988 during a heat wave and drought he said, "with 99% confidence, we can state that the warming during this time period is a real warming trend." Naturally, several scientists are critical of him for showing a lack of discretion, and have publicly rebuked him.

Another group of scientists work with interest groups, have Ph.D.s and are working for a cause to give credibility to the positions of a group. A number of interest groups support research staff, prepare reports and give testimony to Congress. Some prominent interest group scientists are George Woodwell on the board of the Natural Resources Defense Council and trustee of the World Wildlife Fund, Michael Oppenheimer, atmospheric physicist who is a scientist for the Environmental Defense Fund, and Irving Mintzer, Ph.D. from the Berkeley energy program working with the World Resources Institute. Mintzer does policy analysis and examines possible scenarios resulting from different policy alternatives.

POTENTIAL EFFECTS OF THE GREENHOUSE

Dr. Wallace S. Broeker, a professor at Columbia University in New York City warns, "The inhabitants of the planet Earth are quietly conducting a gigantic environmental experiment. So vast and so sweeping will be the impact of this experiment, were it brought before any responsible council for approval, it would be firmly rejected as having potentially dangerous consequences."

The place of the greenhouse scientist remains to issue warnings of future effect on climate, so the general populace can modify their behavior to reduce the forthcoming calamitous effect. Since it will be virtually impossible to stop the carbon dioxide buildup in the atmosphere because of fossil fuel burning and devastation of tropical forests, it becomes a necessity to slow down the process by developing new energy policies, agricultural systems, new human settlements and coastal locations. Different computer models give varying figures for the warming trend, but the National Center for Atmospheric Research (NCAR) in Boulder, Colorado and the United Nations Environment Program estimate an average worldwide temperature rise of 1.5° to 5° C by the year 2030.

Potential effects of the greenhouse effect include rising temperatures, shifting rainfall patterns, more hurricanes and storms, shifting ocean currents and rising sea levels. All these changes will affect the social, political and economic structure of the planet by changing patterns of energy use, pollution, agriculture, forests, fisheries, cities and endangered species. In order to understand how scientists come to predict future climate weather patterns, it's necessary to understand computer modeling.

Climatologists first simulate the current conditions, and then double the amount of carbon dioxide in the atmosphere. The computer runs until the conditions are stabilized, and then a map is drawn to show the change in temperature, precipitation and other factors. The best climate models must oversimplify the complexity of the real atmosphere. Variables that are difficult to estimate are the effect of vegetation, ice and snow, soil moisture and the effect of clouds.

The National Center for Atmospheric Research uses 300 by 300 mile grid boxes representing 4.5 degrees latitude and 7.5 degrees longitude, and treats them as uniform masses of air. Effectively, this divides the earth into a grid of 1,920 squares with nine levels making 17,280 boxes. The weather variables are calculated every 30 minutes and it takes 10 hours on a supercomputer to compute a years worth of weather. Each box is linked to its neighbors, so it can respond to changing conditions with changes of its own.

The Institute of Applied Systems Analysis in Austria did a calculation of climate change for specific nations by using computer modeling for the United Nations Environment Program. Some nations will benefit by the changes in climate, while others will be affected adversely. The largest effects extend to Canada's Saskatchewan Province with a temperature rise of 3.4° C and increased rainfall to 18%. The increased rain is expected to lower the wheat production by 25% and have a knockout effect on the economy. The farm incomes will drop 26%, employment will decrease 1.9% and GDP fall by 12%.

Conversely, both the Central European area of the Soviet Union and Iceland will benefit greatly. The Soviet Union with a temperature rise of 1.5° C will increase wheat yield 30% and increase the farmland suitable for wheat culture. Iceland's 3.9° C temperature promotes a 15% increase in precipitation with increased hay yields and more rainfall with more rangeland and grassland.

A continuing problem affecting the amount of carbon dioxide in the air is the rapid deforestation and removal of trees which act as sinks to remove carbon dioxide from the air. An important activist group demonstrates what can be done in India. Chipko arose as a result of grassroots action by impoverished villagers. As a timber company headed for the woods above the village to remove more timber, the local men, women and children rushed to "chipko" (hug or cling to) the trees daring the loggers to push the axes into their backs.

The Chipko movement has become the most acclaimed forestry movement and continues to progress forward. The villagers are now going from resource protection to ecological management and restoration. From protecting trees against loggers, they now plant trees, build soil-retention walls and prepare village forestry plans. (Worldwatch Paper 88, "Action at the Grassroots").

SEA LEVELS AND COASTAL RISE

As the earth continues to warm, more snow and ice will melt, especially in the polar area which reflects less solar radiation and will warm more rapidly. The largest potential polar effect will occur in the Arctic region where ice at the edge of the Arctic Ocean could melt and never reform. Because of the huge Antarctic land mass with ice thousands of meters deep, the effect will be less there than at the Arctic. If the greenhouse heating progressed, the world's oceans could expand solely from the effect of the rising temperature. A global warming of 1.5° to 5.5° C can cause a sea level rise of between 20-65 centimeters.

Several scientists are studying the effect of sea level rise on different areas of the world, especially low coral islands, coastal regions of continents, river deltas and other islands. Prominent scientists engaged in these studies are Robert Buddemeier of Lawrence Livermore National Laboratories studying the impact of sea level rise on coral reefs, James Broadus and associates at Woods Hole Oceanographic Institution and Stephen Leatherman of the University of Maryland. Buddemeier comments, "If we went all out to slow the warming trend, we might stall sea level rise at three to six feet. But that's the very best you could hope for."

Dr. Stephen Leatherman, Director of the Laboratory for Coastal Research and Director of the Center for Global Change at the University of Maryland, is an authority on sea level rise and its impacts, and has worked on the effects of sea level rise in coastal cities and barrier islands. He assists scientists in Brazil, China, Egypt and India to develop their own beach erosion and land loss assessments. Dr. Leatherman's team at the Center for Global Change predicts a rise in sea level of three feet will threaten fifteen of the world's largest cities. Worldwide sea levels have risen one-half foot in the past century, but the actual rise of water has been over twice that amount for coastal cities because of land subsidence. Much of the world's population lives in a coastal zone with many buildings built at elevation of less than 10 feet above the high tide level. Additionally, sea level rise induces erosion but estimates are that 90% of the U.S.'s beaches are eroding and vast areas of coastal marshlands are being lost in Louisiana at accelerating rates.

Barrier islands along the U.S. Atlantic and Gulf Coasts are at risk, as well as island nations such as Indonesia and the Philippines and low lying island nations like the Maldives. Delta areas are rapidly changing from sediment buildup from the rivers and seas. A rapid rise in sea level will upset this system. Many deltas are densely populated and include cities worldwide, such as Calcutta, Shanghai, Bangkok, Tokyo, Osaka, London, Rotterdam, Venice and New Orleans. Beachfront properties are some of the most valuable real estate in the country and include important cities in the U.S. built on barrier beaches (Galveston, Texas, Miami Beach, Florida, Ocean City, Maryland and Atlantic City, New Jersey).

Pier Vellinga, Ministry of Housing, Physical Planning and Environment of the Netherlands and an associate of Dr. Leatherman, offers some interesting observations on his country's fight with the sea and how they have coped with these problems. In response to a long-term coastal protection plan for the Netherlands, a plan was developed. In 1953, a disastrous storm surge killed 2000 people and flooded a significant part of the Netherlands.

The Dutch believe it is economically feasible to protect large coastal cites such as Rotterdam. They have erected the Rotterdam waterways to protect the city and surrounding area from flooding during storm surges at a cost of over $1 billion. Large scale protection for farmlands are utilized. Some areas are raised to create high land above flood levels, while other areas could be protected by dikes.

How can shoreline erosion be dealt with? The populations can retreat from the shore, armor the coast, or nourish the beach, but the eventual choice will depend on socioeconomic and environmental conditions. If an area is highly urbanized, such as Miami Beach, Florida or Ocean City, Maryland, it isn't realistic to abandon property. Beach restoration can be accomplished or sea walls built to stabilize the shore and provide protection from hurricanes.

Dr. Leatherman stresses that human induced climate change and the resulting sea-level rise are global issues. Although the greenhouse effect is caused mostly by industrial countries, developing countries may well suffer the most. Discussions on the global approach to problems of the greenhouse effect have been initiated by UNEP (United Nation Environment Program). If no restrictions are put on the emissions in developing countries, the current developing countries will be responsible for more than 50% of greenhouse gas emissions by the next century. Global sharing of costs and efforts to limit climate changes and their causes are now being discussed by the international community.

Closely connected with the prospective sea level rise of the greenhouse effect is the issue of oceans absorbing excess carbon dioxide, and the effect of ocean currents. Wallace Broecker, an oceanographer at Columbia University, has become quite involved with both of these facets of oceanic effects. He agrees with James Lovelock that plankton give off chemicals that affect the formation of clouds. These plankton, also, absorb carbon dioxide during photosynthesis and eventually die, sink to the bottom of the sea and entomb the carbon dioxide. At the onset of past glacial periods, the polar ice caps grew and sea levels rose effectively concentrating nutrients in the water. The population explosion of plankton increased the carbon dioxide removal from the water, dropped the oceanic carbon dioxide and pulled additional carbon dioxide out of the atmosphere. This resulted in a cooling of the earth.

Broecker believes oceans may well be the real force of climatic change, since they cover 70% of the planet. They have the capacity to warm up and cool down more slowly than the land, and absorb one-half of all the carbon dioxide released by human activities. Ocean currents effects are important on weather as shown by the Gulf Stream which warms Great Britain, and the effects of El Niño current that appears periodically off the coast of Peru. The El Niño spreads warm water and ruins anchovy fishing for fishermen in Peru wrecking their livelihood.

If the Gulf Stream stopped flowing northeastward and swung south into the mid-Atlantic, it could leave northwestern Europe with a cold climate, pushing the North Atlantic herring and 20 odd species of fish threatened with overfishing, into extinction. The Gulf Stream acts as part of a conveyer belt current containing twenty times as much water as earth's rivers. It carries heat from the Atlantic to the Pacific and from the pole to the equator. This current flows in the North Atlantic, surrenders its heat, becomes cold and dense and sinks. The effects of this current could increase incidence of sea storms including hurricanes.

Perhaps the most vocal of the greenhouse scientists is Stephen Schneider with his frequent Congressional testimony. One representative said to him, "You mean to tell me that you guys have spent a billion dollars of the taxpayer's money proving that the winter is cold and the summer is hot?"

"Yes, sir," I replied, "and we are very proud of it— for if we couldn't reproduce that very large seasonal climate signal in our models, then I wouldn't have the nerve to stand before you and suggest that there is a good likelihood of major climate change from human pollution."

Schneider wrote about speaking at an AAAS (American Association for the Advancement of Science) meeting on the effect human activities have on changing the climate. "After speaking for half an hour or so on how various kinds of human activities could change the climate, I concluded unfortunately, only a relatively few people were even aware of the possibilities."

I then quipped, "Nowadays everybody's doing something about the weather, but nobody is talking about it."

He notes that a distinguished science writer of the New York Times was taking notes of his conversation. He evidently loved one-liners that boil down complex issues or reflect or create controversy. From that time on, Schneider's opinions were no longer his property. A colleague of Schneider's, Roger Revelle calls Schneider "the smartest guy in the climate modeling field."

Climate models differ on the effect in monsoon dependent lands such as Africa or India. However, monsoon rainfall might increase and be a supposed advantage in India. At the same time, frequent monsoons might cause disastrous flooding as occurred in 1988 in Bangladesh destroying farmlands, animal habitats and residences.

At a Senate Energy Committee hearing, Senator Tim Wirth of Colorado introduced a bill to curb the emissions of greenhouse gases, especially carbon dioxide. The Department of Energy argued about the uncertainty of climate change, and the folly of investing in resources when the problem would clear eventually. Why should people pay extra money to insulate their homes or drive smaller more fuel efficient cars to produce less carbon dioxide producing fuel?

Schneider replied, "We can't be certain that the warm and dry summer of 1988 was a result of the greenhouse effect, nor the unusual summer heat in the southeast in 1986 to the greenhouse effect, nor the devastation from hot weather in the Corn belt in August 1983, nor the temperature that killed not only crops but people in the late spring and summer of 1980 in Texas, Arkansas, Oklahoma and Missouri. The hot weather could be random, but the hottest years of the 20th century occurred in the 1980's—in 1983, 1987 and 1988."

At the same hearing, Senator Bradley asked , "I presume that you can tell me what some of those surprises are that might be lurking?"

Schneider replied, "A surprise is by definition something you doubt is going to change."

Some changes will be in the distribution of forest species, unusual changes in ocean temperatures, shifts in growing seasons, new agricultural zones and major changes in the likelihood of forest fires.

The greenhouse scientists enjoy showing the probabilities of climate change by loading climatic dice.

Schneider remarks, "In Washington, D.C. the probability of five or more days in a row in July with afternoon temperatures greater than 95° F is now about one in six— the odds of getting one face of an unloaded die. If the temperatures increased by only 3° F, and nothing else in the climate changed, then the odds of an unpleasant heat wave go up to nearly one in two or three faces of the die loaded by our actions. In Des Moines, the equivalent odds go from one in seventeen to one in five. In Dallas, the odds of five or more 100 plus degrees F days in a row go from one in three to two in three if July average temperatures go up by 3° F.

Schneider uses the win or lose philosophy on the agricultural effect of the greenhouse phenomenon. How might the winners be charged and the losers compensated? If the Cornbelt of the United States were to move north and east by several hundred kilometers from global warming, then a billion dollars lost in Iowa farms could become Minnesota's billion dollar gain. Nations to benefit from the change to a warmer climate Canada and the Soviet Union, who today have much of their land too cold for large scale crop cultivation.

The greenhouse effect is not necessarily bad news for agriculture, because more carbon dioxide in the air means a rise in the rate of photosynthesis and growth. Also, with the increased carbon dioxide, some plants will use water more efficiently and decrease the need for more water. Cynthia Rosenzweig, a NASA Goddard

Institute researcher, has worked extensively with computer models in an attempt to predict the effect of carbon dioxide buildup and climate change on wheat. By using the computer with a doubled carbon dioxide level, she calculated conditions under dry and wet years. When the rainfall is normal the wheat grew better, but in dry years there was an increased crop failure due to excess heat.

The greenhouse effect caused by burning of fossil fuels will affect the lower atmosphere, and will affect hydrologic processes and the location of rainfall. Warmer air will cause more evaporation of ocean water, more clouds and overall rise in rain and snow of 5-7%. Central India could have doubled precipitation, while the centers of continents and middle latitudes such as the U.S. midwest could have drier summers. Very arid areas, southern California and Morocco could have drier winters which will result in drier soils.

In the United States, there could be internal fights for water similar to the ongoing fight for the Colorado River water rights. In California, 42% of the water arises from the Sacramento and San Joaquin River basin fed by runoff from the Sierra Nevada and other mountains. The majority of the water falls as snow in the winter which melts to feed the rivers, reservoirs and subterranean aquifers. A higher temperature in the Sacramento River Basin would mean the precipitation would fall as rain, not snow, and would cause run off. Even with the same total runoff, there would be more flooding in the winter and less water in the summer.

Besides the changes in precipitation and water distribution, the greenhouse effect is fluctuating due to human activity. Humans love to modify the land by deforestation, overgrazing, and woodcutting resulting in desertification. These activities cause the surface to absorb less sunlight and become cooler promoting changes in the lower atmosphere. Cool air arises at the earth's surface and warms air above reducing the convection activity in the atmosphere to suppress cloud formation and precipitation.

Desertification can affect rainfall by increasing dust in the lower atmosphere. This dust absorbs and scatters the sunlight and warms the upper part of the dust layer to prevent some of the solar radiation from reaching the earth. Subsequently, some of the rainfall is reduced.

Environmental threats are now fostering international tension and instability between nations. Egypt, the Sudan and Ethiopia are in the midst of water-supply arguments prompting a foreign minister to warn: "The next war in our region will be over the the waters of the Nile, not politics." Widespread drought in sub-Saharan Africa occurs because the populace is engaged in agriculture affected by rainfall. Some of the on-going effects of the drought are dried-up watering places, withering crops and reduced forage for livestock. The whole social and economic structure of nations is affected as prices rise, some food is imported, malnutrition occurs and surges of people migrate to the countryside to urban centers to refugee camps.

In Dhandhuka on the barren coastal plain of India's Gujarat state, a grassroots movement is taking place to overcome desertification with its social and economic effects. Children became malnourished as their milk cows died of thirst and couldn't provide any more milk, and four-fifths of the population had to migrate because of lack of water. In 1981, the women organized a permanent reservoir to trap the seasonal rains. After observing irrigation channels lined with plastic sheet in other locations, the community agreed to construct a reservoir built in this fashion. In 1986, they moved thousands of tons of earth by hand and completed the pool during the dry season. After the seasonal rains, they were well supplied with water.

EFFECTS OF GLOBAL WARMING ON PLANTS AND ANIMALS

A few degrees change in the climate caused by global warming can make a difference between a species surviving or becoming extinct. Although tropical plant life is still being destroyed by deforestation, a greater impact will be on the species in cold temperate regions and polar regions. As the temperature gets warmer, the plants and animals will migrate northward to the poles. Scientists such as Norman Myers, Ph.D. of Oxford, England try to predict the migration of species by events of the last ice age, when the glaciers retreated and the trees and plants followed northward. They traveled only 25 miles each one hundred years, but with the arrival of the greenhouse effect, the plants and animals will need to travel 10 times as fast, and many will find it impossible to adapt.

Changing climates can alter the mix of plants and animals, and would affect the species potentially that adapt best to increased carbon dioxide, warmer temperatures and a higher rainfall. If rangelands became much wetter, water-sensitive species would start to predominate. Many life cycles of plants would accelerate changing the production of flowers, seeds and fruits to different times. In semi-arid areas where growth is controlled by fires, the fuel might accumulate more quickly with more fires. For example, the 30-40 year fire cycle in San Diego County might accelerate by 5-10 years with more frequent forest fires.

Norman Myers has extensive experience as a Consultant on the Environment for the World Bank, United Nations Agencies and was a Senior Fellow for the World Wildlife Fund. Although born and raised in England, he worked in Kenya for 24 years. At a lecture at the University of Arizona to a biological audience, he remarked, "We have a problem with acid rain, loss of crop soil and desertification. These are all reversible with time and money, but will take billions and trillions of dollars. The better news is the cataclysm of extinction hasn't arrived and is not inevitable. The best news is that all people are beginning to sit up and take notice."

Norman Myers appears very intense and dedicated to saving the planet's species. Dressed conservatively in a navy jacket and gray slacks, he appears to be the prototypic new scientist with his dark beard and glasses. Everyone applauded enthusiastically at his remarks, and acknowledged his contributions to saving planetary species. He continued his remarks. "We should pause and stop extinction during the decade of the 1990s. If a species becomes extinct, it is gone for good. At your neighborhood pharmacy, the chances are one in two that your analgesics, laxatives and diuretics come from plants, many of which are found in the tropical rain forests. For example, the rosy periwinkle of an endangered species in Madagascar has two alkaloids which are effective against leukemia and Hodgkins Disease. Through slash and burn agriculture, forest destruction has doubled and very little remains of the forest in Brazil. Acid rain has depleted the forests in China. We act as though we have a spare planet to spare. We can't let these creatures disappear in such vast numbers. We should be like Prince Charles with his optimistic outlook and have faith and keep hoping."

He told the audience that the Chipko movement in India took three years before becoming effective. These disadvantaged people started a grassroots protest by hugging the trees and refusing to leave until the loggers left. Myers admonished the audience to support their favorite conservation groups, write to Washington, D.C., especially environmental congressional people, do an environmental audit of their lifestyle and start recycling materials.

Edward 0. Wilson, Ph.D. of Harvard, where he is now Professor of Science and Curator of Entomology agrees with Norman Myers on the threats to biodiversity. The biota are being threatened equally by deforestation and climatic warming. The habitat loss is more apparent in tropical rain forests, but there will be a great effect on the biota of cold temperate areas and polar regions. The greatest effects will be on plants which are immobile and lack the ability to disperse. University of Minnesota scientists Margaret Davis and Catherine Zabinski predict that global warming will affect four North American trees—yellow birch, beech, sugar maple and hemlock and will displace them northward 500-1,000 kilometers. Not only the trees, but others will be displaced with potential to become extinct.

Wilson believes steps can be taken to halt the extinction process by mapping the world species and identifying places that need more conservation efforts. Conservation should go hand-in-hand with economic development, especially in countries where poverty and high population densities predominate. He recommends sustained harvesting of forests rather than clear cutting for agriculture and timber. After tropical forests are clear cut for agricultural purposes, the topsoil is so nutrient poor that agriculture can't take place any longer in that location.

Under cloudless blue skies and amid creosote bushes and cacti stands one of the most important environmental experiments of all time. Biosphere II, a 3.15 acre, airtight experiment with an airtight glass and metal structure is being constructed on a hillside at the 2,500 acre Sun Space Ranch, 30 miles north of Tucson in Oracle. In September 1991, eight hand-picked scientists plan to enter the biosphere and become sealed for two years into this complete module.

Carl Hodges, Director of Environmental Research Laboratory at the University of Arizona remarks, "We've got acid rain problems, dirty air problems, and we've got carbon dioxide increasing at an alarming rate. Right now, we don't have a research tool where we can control the global parameters, like carbon dioxide and the quality of the atmosphere. I see the big payoff of Biosphere II as learning how to do a better job of stewardship of Biosphere I (the biosphere of the Earth.)"

On March 8, 1989, Abigail Alling, a 29 year-old marine biologist, spent five days inside a terrarium-like environment in a Biosphere II experiment. She was part of a larger experiment where all air, water and nutrients were recycled and isolated from the outside. She commented, "The three aspects of science: The experimenter and the experience became a unit. This is really a historic day. It's truly a paradise. The tropical garden and the whole test module...I was in some ways very sad to come out."

The Biosphere II project was conceived by the Institute of Ecotechnics in London and paid for mostly by Texas centimillionaire Ed Bass at a cost of $30 million: it is known as Space Biosphere Ventures (SBV). The 3.15 acre, 7 million cubic foot mini-Earth has seven distinct biomes: savanna, marsh, desert, tropical rain forest, a 35 foot deep ocean, intensive agriculture and human habitat.

This Star Trek-like enterprise may be instrumental to future space pioneers who work on Mars or on the moon. In such a hostile environment, people, plants and animals could survive in an airtight biosphere. Scientists plan to conduct research on plants, animals, air and water to discover how the earth works and how we might be better stewards of planet Earth.

What is a biosphere? It's an enclosed, ecological system and is a relatively complex evolving system within which flora and fauna support themselves and renew their species, consuming energy. Earth supports a biosphere. NASA has done experiments on Closed Ecological Life Support systems, but is less advanced than SBV. NASA is trying mechanical systems for purifying air and water instead of biological systems. The U.S.S.R. has been working on a modest biosphere project, but is more interested in studies on the effects of long-term human isolation and uses artificial systems to purify the air and water.

The atmosphere of Biosphere II is a continuous system and circulation is engineered by convection, mostly. The flow of air is accomplished by trade winds driven from the desert to the rain forest by the sun's heat. Upon cooling by condensers, the air will release moisture over the forest, and then be blown back to the desert by fan. In this closed system, the plants give off oxygen and carbon dioxide from animals is recycled to plants.

Representative Claudine Schneider (R-RI) works to improve aspects of the current greenhouse effect. The World Wildlife Fund and The Conservation Foundation told of her efforts as a member of the Subcommittee on Natural Resources, Agricultural Research and Environment and how she recently introduced a a Global Warming Prevention Act (HR 1078). She suggests the following points to alleviate global warming.

1. Revise the Montreal Protocol to phase out CFCs and other damaging halocarbons by 1995, rather than the current schedule of 50% phaseout by 2000.

2. Establish national and international goals to reduce carbon dioxide emission 20% below current levels by the year 2000.

3. Support an international agreement on population growth to achieve universal family planning.

4. Tax incentives for consumers to purchase fuel efficient cars, increase the gas guzzler tax on grossly inefficient cars and increase national fuel economy standards 75% over the next decade to 45 mpg.

5. Increase R&D funding for energy efficiency and solar/renewable energy resources which have been cut 50% to 75% respectively over the past decade.

6. Promote "least cost" energy planning, domestically and internationally, to spur energy efficient improvements.

7. Slow deforestation and promote reforestation in U.S. where only one tree is now replanted for every four that die or are cut down.

8. Helping tropical forest-rich countries implement forest management and conservation plans.

9. Allow up to half of debt owed to the U.S. Government to be reduced when applied to specific resource preservation and energy.

10. Promotion of nationwide waste reduction and recycling programs.

The next two chapters discuss important facets of the world greenhouse effect, and their contributions to the overwhelming problems. It's commonly known that CFCs contribute to the ozone depletion, and allow ultra violet radiation to enter our atmosphere in ever-increasing amounts. The ozone depletion is a relative newcomer on the scene, and involves many scientific expeditions to the polar regions by very adept scientists. CFCs are important as greenhouse gases.

Citizens are being advised that the solution to the greenhouse effect lies in the adoption of more nuclear plants. Doctors, scientists and activist groups help the general public to realize the danger of further adoption more nuclear power plants, and warn of the concurrent dangers of nuclear energy testing.

5 MISSING SKY SCIENTISTS

"A dramatic reduction in the vital ozone layer over Antarctica proves the greenhouse effect is real and presages a gradual warming of the Earth that would raise ocean levels, erode coastal cities, bring dustbowl conditions, increase human misery and if not checked, lead to eventual extinction of the human species, scientists warn Tuesday." (Associated Press June, 1986)

Warnings about the ozone layer depletion over the Antarctica first arose in an article published by Sherry Rowland, Ph.D. in the international scientific journal, "Nature" in June of 1974. Sherry Rowland first received a glimmer that things were terribly wrong with the stratosphere when he attended a meeting of chemists discussing atmospheric chemistry in 1972. The next year, he obtained official permission from the Atomic Energy Commission to study CFCs and acquired a postdoctoral fellow, Mario Molina, to assist him with his CFC studies at the University of California in Irvine. Their experiments with CFCs showed remarkable chemical compounds: they don't interact with living things, don't dissolve in the oceans and don't incorporate into rain, but eventually end their life in the stratosphere. By the fall of 1973, the team of Rowland and Molina knew the CFCs would reach the stratosphere, be broken apart by ultra violet radiation and eventually release chlorine atoms. A single chlorine atom can scavenge and destroy many thousands of ozone molecules, and Rowland's and Molina's conclusions appearing in "Nature" were summarized by this abstract.

"Chlorofluoromethanes are being added to the environment in steadily increasing amounts. These compounds are chemically inert and may remain in the atmosphere for 40-150 years, and concentrations can be expected to reach 10-30 times present levels. Photodissociation of the chlorofluoromethane in the stratosphere produces significant amounts of chlorine atoms, and leads to the destruction of stratospheric ozone."

Years later at a Senate Hearing in 1986, Rowland added these comments about their discovery. "In early 1974, no measurements had yet been made of any chlorine-containing molecules anywhere in the stratosphere. Now, we have detailed evidence concerning at least ten chlorinated compounds in the stratosphere itself, and many more throughout the lower atmosphere. The chlorofluoromethanes CCl_3F_2 (Fluorocarbon-120) are, as they were designed to be, chemically inert...The absence of effective removal processes for these chlorofluoromethanes led us to predict two important consequences: the average molecule of each would survive in the atmosphere unchanged for many decades: and removal would occur by destruction in the stratosphere after absorption of ultraviolet radiation."

Rowland mentions that atmospheric measurements show the accumulation of CFCs in the lower atmosphere at a rapid rate, and they are now found at almost three times the concentration measured in the early 1970's. The average atmospheric lifetime remains long, and is placed at 70 years for Fluorocarbon-11, and more than 100 years for Fluorocarbon-12. Scientifically, these measurements were made by capturing stratospheric air in flasks, and returning the flask to the Earth before opening them by remote control.

Rowland fears that underdeveloped countries will soon add to the CFC overload of the atmosphere and stratosphere and made these remarks at Chapel Hill in March, 1987, "Any agreement which affects only North America, Western Europe and Japan will soon be overwhelmed by the increased usage outside the regions....."

The ozone missing sky remains an ongoing problem that must be resolved by governments throughout the world, as well as environmental organizations and scientists. William Walsh, an attorney for the U.S. Public Interest Group, says the depletion of the ozone layer is, with the possible exception of nuclear weapons, one of the gravest crises we face here on Earth. A history of the movement to ban the troublesome CFCs shows increasing concern by a number of nations, the United Nations, and by groups of cooperating scientists from different countries.

As early as 1978, the U.S. Environmental Protection Agency (EPA) and the Food and Drug Administration banned the use of CFCs in non-essential aerosols. However, since this time the non-essential aerosols application has ballooned until the CFC levels rose beyond the pre-aerosol ban levels. As time marched on to 1982, the U.N. Environment Program (UNEP) sponsored negotiations trying to develop international agreements to protect the ozone layer. The Vienna Convention for the Protection of the Ozone Layer in 1985, and the Montreal Protocol were outcomes of the United Nations interest in the ozone depletion problem.

The U.N.'s interest became more urgent when a group of British scientists led by Joseph Farman found a 40% drop in the springtime ozone level high above the Antarctic and published their results in 1985. Fear and apprehension about the use of CFCs arose to an all-time high when the British Antarctic Survey alerted the nations of their findings. Farman's team first noticed a hole in the sky in Halley Bay, Antarctica in the winter of 1981. Farman had been measuring ozone levels since 1955, but didn't notice a decline in levels until 1977. They had been using a spectrophotometer which measures the concentration of the atmosphere by taking spectroscopic fingerprints. Their instrument was fairly old, so they bought a new one to use in the measurements.

Using the new instrument, Farman's team remeasured the depletion of the ozone hole over Halley Bay and reported in October, 1984 that the ozone depletion amounted to 30%. The results of their finding appeared in the "Nature" journal on May 16, 1985 and caused much apprehension at the Goddard Space Flight Center responsible for the failure of the Nimbus 7 instruments which measured ozone levels. The computers had been programmed to reject any data lower than 180 Dobson units as an erroneous measurement, but fortunately the measurements had been saved even though they appeared to be wrong.

After the new measurements appeared, the NASA scientists reprogrammed the computers to take the measurement data and stop rejecting the results. The ozone hole extended over the whole Antarctic continent, and the depletion occurred in a layer of atmosphere 6 and 15 miles altitude. After Farman's warning, scientists felt a global emergency existed, and everyone rushed to study the Ozone Hole in the Antarctic. Susan Solomon of the National Oceanic and Atmospheric Administration led the first National Ozone Expedition (NOZE) to McMurdo Station in Antarctica. When her expedition landed at McMurdo, the amount of equipment was staggering, and included 15,000 pounds of oscilloscopes, sensors, tape drives and tanks of liquid nitrogen with dozens of high tech instruments.

Solomon evolved the hypothesis of the mechanism of ozone layer depletion involving clouds in the stratosphere furnishing surfaces of ice to catalyze chemical reactions. She also evolved an explanation of the place of nitrogen oxides in the over-all chemical reaction. Nitrogen oxides could convert to nitric acid, the acid could provide a surface to catalyze chlorine's attack on the ozone, and subsequently remove nitrogen from circulation. Nitrogen remains an element that could stop chlorine from devouring ozone.

Her expedition used instruments aboard a giant helium balloon, raised aloft three times a day. This expedition found 100 times the chlorine in the stratosphere than normal. Also, the volcanic effects of the volcano El Chichon that erupted three years ago were not currently present as an effect on the findings and blew a prevalent volcanic effect theory. Shocked, this expedition found an ozone hole larger than any that had ever existed.

The next expedition in August, 1987 included scientists and personnel from nineteen organizations and four nations who met in Punta Arenas, Chile to conduct the Airborne Antarctic Ozone Experiment. By 1987, the ozone hole was the largest ever measured with current measurements of a combination of ground and satellite based measurements. Some individual scientists from this expedition will be spotlighted.

The ozone scenario changed to include two new scientists responsible for even more exciting conclusions on the ozone depletion problem! Bob Watson became the NASA scientist in charge of the Punta Arenas mission that occurred in August, 1987. A fellow scientist, James Anderson, an atmospheric chemist from Harvard, held the responsibility for the precise measurements of chemicals in the stratosphere. Anderson, a highly respected scientist, is soft-spoken and very intense in his work. The measurements Anderson made were difficult because the stratosphere is far distant, and the chemicals to be measured were present in very small amounts.

Anderson's experiments required him to determine how much chlorine floats in the Antarctic stratosphere, and to find whether this amount accounts for the rapid breakdown of ozone. He measured the rate of reaction by monitoring levels of chlorine monoxide, an unstable by-product of the collision of ozone with chlorine. Every piece of his equipment was hand built by his team, and absolutely nothing was left to chance.

Watson said,"The smoking gun will be chlorine monoxide. We have to find out how much there is in relation to other chlorine species." How was the data collected from this highly technical experiment? It was collected in the ER-2, a research version of Lockheed's famous U-2 spy plane and the best aircraft for high altitude flights. The scientific instruments came packaged in 2 pods, and the planes flew at an elevation of 12.5 miles where balloon flights showed the most ozone depletion. The ER-2 made twelve passes through the Antarctic stratosphere in August and September. The results showed that there was a steady rise in chlorine monoxide until the levels were more than 500 times the normal concentration. At the same times, the ozone levels dropped steeply. The ozone hole appeared deeper this year than any other year.

Anderson remarked, "What we saw in the Antarctic was a clear connection between high chlorine levels and ozone depletion.... It's time we stopped using the atmosphere as a test tube for global chemical experiments." Watson concurred with Anderson, "if we find the ozone is decreasing in Antarctica and that it is a precursor to global ozone depletion, then we will have to find a way to immediately get rid of CFCs for all but the most essential applications, such as refrigeration."

Bob Watson gave the conclusions at a press conference in Washington, D.C. September 30, 1987. After six weeks of flying over 110,000 miles, the region of the ozone hole is low in water and in nitrogen compounds, while every chemical involved in ozone is present in abnormal concentrations. There is a close correlation between measured concentration of ozone and chlorine monoxide. Ozone is the least where chlorine monoxide is the greatest.

In the fall of 1987, 24 nations met in Montreal, Canada to limit harmful CFCs and halons, but recently the agreement has been found to be insufficient and limiting in safeguarding the ozone shield. The hole over the Antarctic is expected to increase and would require strict controls amounting to an almost total ban of CFCs in order to redress deficiencies. Even with the international ban the ozone will not be replaced for centuries.

Farman and the new ozone scientists met several years after the first interest in the ozone layer at the Dahlem Workshop in Berlin in November, 1987. The scientists included Lovelock, Rowland, Crutzen, Farman, Philip Solomon and Watson. At the workshop, they established that chlorine from industrial products, chiefly CFCs is to blame for the destruction of ozone in the Antarctic. Farman and colleagues agreed that the Montreal agreement committed us to increase increased stratospheric chlorine levels by the year 2020. This is ten times the concentration of chlorine in the atmosphere before the use of CFCs became widespread.

Several new international agreements developed in March and April of 1989. The twelve European Economic Community nations (EEC) agreed to eliminate the production and use of ozone destroying chemicals by the end of the century. Those chemicals involved were CFCs used as coolants, propellants, insulating material and solvents. Another 123 countries met two days later in London at a Saving the Ozone Layer Conference sponsored by British Prime Minister Margaret Thatcher. Twenty nations agreed to ratify the the Montreal Protocol; among them were several developing countries, Brazil, Philippines, Malaysia and Turkey. Even China seriously considered agreeing to ratification.

Another meeting concerning the ozone layer occurred in Helsinki, Finland when 80 nations met in April, 1989 and agreed to strengthen the 1987 Montreal Protocol. These nations agreed to phase out the production and consumption of CFCs as soon as possible and not later than the year 2000. They concurred that the Protocol Agreement calling for a 50% reduction by 1999 was out of date. They called for a phaseout of halons which are ozone destroying chemicals used in fire extinguishers. One omission in the agreement involved a lack of reduction in carbon tetrachloride and methyl chloroform. The EPA mentions that unrestricted use of these chemicals could double chlorine contamination in the ozone layer in 90 years.

POLAR OZONE HOLES

What has happened to the ozone hole two years later in 1989? A huge surprise is a severe ozone depletion comparable to the 1987 ozone hole. David J. Hofmann, physicist from the University of Wyoming at Laramie says, "We weren't expecting it to be as severe as 1987. That was a surprise. This year and 1987 were close to as bad as you can get."

The ozone disappeared completely from certain stratosphere layers. Hofmann remarks, "We're approaching total depletion in the 16 to 18 km range. In that slab we've gone from 50 Dobson units (of Ozone) at the end of August to something like 5 Dobson units." Although scientists thought 1987 was an abnormal year, new studies confirm that 1989 may match 1987 and may be more severe.

The ozone depletion covers an area twice the area of the Antarctic continent. The readings of average ozone in 1987 were 250 Dobson units on October 5th, but on October 3rd 1989, the average dropped to 150 Dobson units. During winter darkness, stratospheric temperatures inside the wind vortex drop to -85° C., when nitric acid and water condense to form cloud particles. These particles may play a critical role in chemicals that destroy ozone.

If the ozone depletion is so great over the Antarctic, might there be a problem over the Arctic region? In the winter of 1989, scientists traveled to Norway to find whether there is a threat to Arctic ozone. They used balloons carrying sensors to measure ozone levels and count cloud particles; two expeditions of scientists in both Norway and Sweden found high stratospheric levels of the same chlorine compounds that attack ozone in the Antarctic. David J. Hofmann of the University of Wyoming and his colleagues found a thinning in the Arctic ozone between 22-26 km in the stratosphere similar to the Antarctic. A similarity in both the Antarctic and Arctic findings suggest there may be a beginning of an ozone hole over the Arctic. Because the polar vortex (a wind pattern circling the polar regions in winter) is unstable, the ozone level should not drop as severely as the Antarctic.

WHY IS OZONE A PROBLEM

How do CFCs and halons destroy the ozone layer? These chemicals may remain in the atmosphere for decades, but are relatively inert which means they don't react with other chemicals. CFCs gradually move upward until they reach the stratosphere where they break down to release chlorine with the chlorine causing the eventual ozone destruction. The catalyst causing the breakdown of CFCs and halons is the ultraviolet light in the stratosphere. The chlorine atoms collide with the ozone molecules, and the chlorine grabs ozone's third oxygen molecule. Radicals with a high number of electrons are reactive, and when chlorine monoxide meets a free oxygen atom the oxygen in the chlorine monoxide meets a free oxygen atom: the oxygen in the chlorine monoxide becomes highly attracted to the free atom, and breaks away to form new oxygen molecules. The abandoned chlorine is free to begin the ozone destruction, again. One chlorine atom can destroy 100,000 ozone molecules before is is removed from the atom.

How are CFCs and halons released into the atmosphere? They're released when a product is manufactured, used or disposed of and disposal remains a major source of emissions. If CFCs are emitted at the point of manufacture, the CFCs may be recaptured or recycled, or the manufacturer can substitute different chemicals for CFCs. When products are discarded or serviced, the CFCs would be recovered and recycled, but this isn't happening.

With the advent of our chemical society, there are many products using CFCs. In 1978 most CFC-containing aerosols were banned, but they remained present in a few other related items like boat horns, tire inflators and canned confetti. The major use of CFCs come from coolants, plastic foam and solvents.

CFCs leak during auto air-conditioning service, or when coolants are added to your auto refrigeration unit. It's important to utilize service stations that use "vampire units" to capture and recycle CFCs—many auto dealers have these available and more will have them shortly. When refrigerators are discarded, often the CFCs used as coolants will escape. It's important to ask utility companies to pick up old appliances and dispose of the coolant involved.

Other uses of CFCs include plastic foams such as insulation, mattresses, seat cushions, foam picnic coolers, egg cartons and foam packaging chips. An alternative product to use in saving the ozone layer is the CFC-free Union Carbide Ultracel. Electronic equipment is cleaned frequently with CFC compounds, and the consumer should avoid any products with trichlorofluoroethane prominently displayed on the label.

A constant problem results when some emissions are banked, that is either gradually released, or not released for decades. Some problem appliances that consistently are banked include auto air conditioners, urethane foam, chillers, retail refrigerators, home refrigerators and freezers.

Some new products without CFCs are produced by Cryodynamics, a New Jersey company producing refrigerators cooled with helium rather than CFCs. Petroferm is a small Florida company, who developed a solvent made from fruit rinds to replace solvents made with CFCs. A problem with phasing out CFCs immediately is voiced by Kathy Forte, spokeswoman for DuPont Company in Wilmington, the world's largest producer of CFC's. "We could phase out CFCs tomorrow, but there is $135 billion of equipment dependent on CFCs. If CFCs were banned immediately, there could be disruptions in air-conditioning, food processing, and the manufacture of electrical components. We believe the alternative equipment must be in place before CFCs are phased out."

CANCER AND PLANT EFFECTS OF OZONE DEPLETION

In November 1986, the EPA released a 4 volume report, *Risk Assessment of Stratospheric Ozone Depletion* and concluded, "The United States can expect 40 million additional skin cancer cases and 800,000 deaths for people alive today and those born during the next 88 years if there is no further action taken to limit CFCs." This report warns people of the effect of ozone depletion, but how many are really aware of the problem?

The ozone in the stratosphere helps to absorb visible light from the sun, but light of a shorter wavelength such as ultraviolet light can get through, more so with additional ozone depletion. The three types of ultraviolet light include Ultraviolet-A or UV-A which is not harmful, but Ultraviolet-C and ultraviolet-B (UV-B) can be quite harmful to people's health. Currently, the UV-C is being absorbed by the ozone layer, but UV-B is reaching the ground in a greater degree and is causing quite a few problems. Skin cancer linked to UV-B radiation are cutaneous malignant melanoma and non-melanoma. The EPA calculates that for every 1% decrease in the ozone concentration, there will be a 5% increase in non-malignant skin cancers a year, about 10,000 to 20,000 cases over current figures. Non-malignant cancers can be removed surgically, but the malignant melanomas are more serious. These cancers may develop later in life, since people are living longer lives.

Other health problems linked to increased UV-B radiation are a suppression of the immune system which helps protect us against tumors and other effects such as herpes virus infections, hepatitis and parasite-caused skin infections. Additionally, Emory University at Atlanta, Georgia estimates a 1% decrease in the ozone would increase cataract victims by 25,000.

Besides the potential health effects to humans and animal species, research shows a possibility of many damages to ecosystems, aquatic species and a decrease in agricultural crop productivity. Alan Teramura, Ph.D., the Chairman of the Faculty Advisory Committee of the Center for Global change and Chairman of the Department of Botany at the University of Maryland, is a foremost expert on the effects of ultraviolet radiation on plant life. He's been a guest professor at the Botanisches Institut at the Universitat Karlsruhe in the Federal Republic of Germany in 1982 and Visiting Professor at the University of Hawaii in 1987.

His papers published with colleagues indicate that stratospheric ozone absorbs much of the shortwave solar ultraviolet radiation (UV-B). Therefore, a reduction of the ozone layer will result in an increase in shortwave solar radiation.

Most of the current research available centers on the effects of plant growth and physiology under artificial UV-B irradiation supplied in growth chambers or greenhouses. Teramura found that the damaging affect of UV-B radiation varies among species and varieties of a given species. Specific effects he has observed are reduced growth, reduced photosynthesis and damage to the flowering mechanism resulting in reduced crop yield. The damage to photosynthesis affects photosynthetic enzymes, metabolic pathways, photosynthetic pigments and stomatal function.

The damage observed to different species and varieties can be influenced by the environment. In some experiments, there is a partial inhibition of photosynthesis reflected in reduced carbon dioxide uptake. Soybeans are a plant that shows the effect of UV-B on photosynthesis. Soybeans show a 25% reduction in seed yield when supplemental UV-B corresponding to a simulated 25% reduction of stratospheric ozone is artificially applied by lamps. Damages to the reproductive system involves DNA problems. UV-B radiation absorbed by DNA can lead to heritable DNA lesions. However some higher plants can repair their systems and undo DNA damage. Pollen walls and pollen tubes can be inhibited by UV-B and some germination is inhibited.

PROBLEM AREAS AND ACTION GOALS

The Beacon Journal of Akron, Ohio reported these statistics about CFC-using products in 1988. Of the CFCs used in the U.S., 28% are used in air conditioners and refrigerators and 28% used in plastic foam and insulation. 12 % are used in solvents for cleaning metals, electronic parts and computer chips, 4% in sterilizing medical equipment and 5% in aerosols exempted from the 1978 ban. Another 22% can be found in overseas trade, unreported military and inexact accounting of sales.

Car air conditioners remain the single largest source of the U.S. contribution to ozone depletion according to scientists and testimony before legislative committees. Car air conditioners use

CFC-12, the most damaging of the CFCs. Originally CFCs were developed as a refrigerant in 1930 by General Motors Corporation chemists, and were supposed to be an ideal chemical with nontoxic and non-combustible properties. The car air conditioning problem recently accelerated with 90% of all new cars equipped with air conditioning.

Stephen Anderson of the EPA remarks, "Auto air conditioning represents 20% of America's release of ozone-depleting chemicals." Usually, about half of CFCs are released into the air intentionally during the servicing of auto air conditioners—recycling would eliminate 10% of U.S. releases. When air conditioning is serviced, refrigerants are removed by venting them into the atmosphere and then the unit is recharged. If the venting was eliminated, recovery equipment could be used to reprocess the CFCs to use again.

Other CFC waste comes from small cans of refrigerant sold in stores. Consumers often use these 14-oz cans to charge leaky systems and don't bother to find the leaks. Buyers are urged to avoid using the 14-oz cans, and have their air conditioning leaks fixed. Consumers can urge their auto shop or mechanic to invest in recycling equipment, or they can patronize shops that do.

Besides the problem of auto air conditioners, home and retail refrigerators and freezers account for 6-7% of total CFC emissions. The CFCs used in these appliances are labeled R-12 to distinguish them from CFCs used as blowing agents. The names of the coolants used in refrigerators and freezers are Freon, Isotron and Genetron.

The home refrigerators and freezers consist of an airtight system containing the refrigerant—the compressed R-12 gas is cooled to a liquid and expands to allow evaporation. During the evaporation, R-12 absorbs heat from the unit and releases the heat to the outside. Most of the problem emissions occur when refrigerators and freezers are discarded, serviced, manufactured or when leakage occurs.

Like the auto air conditioning, recovery units known as "vampires" can remove the refrigerant, purify it and recycle the R-12. However, several problems arise in the recovery. In order for the process to be profitable, the salvager needs to know that the cost of the recovered refrigerant is close to or under the virgin refrigerant. Also, the salvager must he willing to recover the refrigerant when disposed. Some new chemical substitutes for R-12 are FC-134A, R-502, mixtures of R-12 with other chlorofluorocarhons or R-22.

Working mothers need help to reduce the time they must spend cooking for their families. Many mothers opt to patronize fast food places such as Wendy's, McDonalds, Kentucky Fried Chicken and Burger King. The American populace marches out of these places with quantities of plastic foam food containers to preserve the freshness and hotness of the prized hamburgers and chicken. I even saw a segment on TV where McDonalds supplied the residents of a nursing home with 100 hamburgers packaged in plastic foam. The fast food chain pledged in the past to stop using CFCs-type packaging, but indications are that he practice hasn't stopped, and that the containers still contain CFCs in the form of HCFC-22, a lesser evil.

Several sources say HCFC-22 releases less ozone-destroying chemical but is still not acceptable. A children's crusade called Kids Against Pollution was started three years ago by a 5th grade civics class at the Tenakill School in Closter, New Jersey.

Over 800 chapters of the kid's group did come forth in the U.S. and Europe to boycott McDonald's stores. Partly due to their efforts and the Environmental Defense Fund, McDonald's began eliminating their sandwich containers which are replaced by paper products. Under pressure from environmental groups, President Edward Rensi said the company had decided "to do what's right." The non-profit Environmental Defense Fund worked with McDonald's since August of 1990 to devise the phase-out. "The clamshell (box) has been a huge symbol of the throwaway society, and we're turning the tide on that," said Fred Krupp of the Environmental Defense fund.

The type of foam used by fast food places is rigid (closed cell) foam, but flexible foam is right behind the rigid foam in frequency of use. The rigid foams are used mostly for packaging, insulation and food service items. Rigid urethane foams are applied in construction, such as sheathing and wall, roofing, basement, floors and refrigeration. Right now, the FDA has approved the use of HCFC-22 for fast food containers, although reports indicate this chemical still destroys the ozone layer. Replacements for rigid urethane foam can be cellulose fiber insulation, fiberglass or non-urethane foams, plastic film bubble wrap, wood shavings and "Evacuated" or "vacuum" panel insulation.

Flexible foams use CO_2 as a blowing agent combined with methyl chloride of CFC-11. This open cell foam can be found in furniture, car seats, bedding, carpet underlay and cushioning. Other chemical substitutes for CFCs could be used such as methylene chloride, CFC-123 and CFC-141b which are low in ozone depleting factors.

Industrial solvents are responsible for a quarter of the total CFC emissions, because such huge quantities are produced and used. The products currently in use are CFC-113 and methyl chloroform used to clean and degrease metal parts and semiconductors, in dry cleaning operations, nonflammable adhesives and pesticide formulas. Petroferm, Inc. of Fernandina Beach, Florida and AT&T (Princeton, New Jersey) have manufactured a new solvent made from citrus fruit rinds called BIOACT(R) FC-7. This solvent does show promise as an answer to CFCs and ozone depletion.

Another new solvent being tested is Genesolv 2010, a trade name for mixed HCFC-123 and HCFC-141b. The EPA announced that a panel of 280 industrial and government reseachers identified this degreaser as a potential CFC-113 substitute with less environmental threat.

Many hospitals started using gas sterilants for their medical instruments instead of the old-fashioned steam sterilization. By using gas mixtures, the use of CFC-12 has increased, and the standard mixture consists of 88% CFC-12 and 12% ethylene oxide by weight. The best over-all solution would be to substitute other chemicals for the CFC-12 used to dilute the ethylene oxide. A problem with the use of ethylene oxide is that it is a suspected carcinogen, and other substitutes are being planned, for instance chlorine dioxide. The Scopas Technology Company of New york has developed new sterilization techniques using chlorine dioxide.

The last problem area lies in the use of fire extinguishers using halons. The halon structure is similar to CFC-11, but uses bromine instead of chlorine and Halon 1211 consists of both chlorine and bromine. Usually halon extinguishers are used for equipment that might be damaged by water or chemicals used in other fire extinguishers. Examples are computer equipment, books or museum pieces, airlines, battle tanks, areas with valuable electrical and telephone equipment, ship engine and boiler rooms. There isn't any problem with home fire extinguishers which use dry chemicals.

Nations are aware of the ozone depletion problem and a recent London conference convened in June 1990. Nearly 100 nations agreed to a ten-year plan to end production of CFCs and other threatening chemicals. Not all of the nations signed the agreement—59 nations signed at the London conference and 30 others acted as observers. This conference improved upon the 1987 Montreal Protocol by agreeing that limitation wasn't adequate, and that halons and CFCs should he phased out completely by the end of the century. Two other chemicals, carbon tetrachloride and methyl chloroform are danger ingredients added to the list to be eliminated by the year 2000 or 2005.

A special fund was approved of $240 million over the next three years to help developing nations switch to more expensive but less harmful chemicals. This fund was held up until the Bush administration agreed to support it. After the Bush reversal, India and China agreed to sign the agreement, and other developing countries have a leeway of 10 years to phase out destructive chemicals.

A recent addition to chemicals depleting the ozone layer is 1, 1, 1-trichloroethane contained in 141 products, from artist's varnishes to hornet-killing sprays, and Scotchguard fabric protectors. David D. Doniger, Director of the Natural Resources Defense Council said, "The most difficult decisions remaining relate to 1, 1, 1-trichloroethane. For most applications, there are alternate products on market that serve the same purpose both in industry and individual homes."

The council said current options under consideration range from a 25% rollback to a complete phaseout in the next 10 years. Japan, the United States and the European Community balk at going further than a 50% reduction. However, the 3M company, makers of Scotchguard, are working on a replacement chemical for 1, 1, 1-trichloroethane.

Past, current and future CFC use is an ever-present threat to the protective stratospheric ozone layer. The United States need to provide leadership and financial help to aid Third World counties to switch to substitutes. Alan S. Miller, Director of the Center for Global Change at the University of Maryland has these comments to make, "with respect to acting to protect the ozone layer, I have been involved in advocating government restrictions on CFCs and other ozone threatening chemicals since the late 1070s. My conclusion from this experience is that the government has a critical role in creating incentives for the development and introduction of substitutes for ozone threatening chemicals. From 1978 to 1986, industry did almost nothing because of the perception that no regulation was forthcoming. Once the Montreal Protocol was signed, industry began to move expeditiously to identify alternatives. I support efforts to phase out 100 percent of these chemicals and expect this will occur by international agreement in London this June."

Government secrecy still shrouds the recent problems of the ozone layer depletion with much of the research conducted by government agencies. Ultraviolet radiation can dramatically increase cancer rates, immune system damage, and cataract formation in eyes. An even more devastating type of radiation, alpha, beta and gamma rays from nuclear energy plants and weapon testing will be dramatized in the next chapter.

6 RADIATION ACTIVISTS

Dr. Linus Pauling, Nobel Laureate in Chemistry and Nobel Peace Prize winner, said after he read *Deadly Deceit: Low Level Radiation, High Level Cover-Up* by Jay Gould and Benjamin Goldman of the Radiation and Public Health Project, "This shocking report on the millions of deaths caused by radiation from nuclear weapon testing and nuclear power plants should be read by every concerned person worried about the health and well being of infants, children, adults and old people."

This new book published in 1990, establishes a link between low-level nuclear emissions and serious health problems which result in an increase in mortality rates. Accidental releases from nuclear power plants and fallout from atmospheric bomb tests can cause frightening increases in infant mortality. The author calculates that 9 million excess deaths occurred from the atmospheric bomb tests of the 50s and early 60s.

One of the most infamous meltdowns that later resulted in governmental denials and concealment of radiation and detrimental health effects began at Three Mile Island on March 28, 1979. Three Mile Island lies in the middle of the Susquehanna River, and is located only ten miles from Harrisburg, the capital of Pennsylvania. The resulting accident had several factors; operator inexperience, faulty reactor design and construction, and an attempted cover-up of the resulting damage to the people and surroundings. According to Daniel Ford, former executive director of the Union of Concerned Scientists, the operators were not qualified nuclear engineers, or college graduates and had not received detailed technical training to handle reactor emergencies.

Unit 2, a Pressurized Water Reactor, built by Babcock and Wilcox, had been operating for less than 3 months, and was riddled with a history of breakdowns. On the night of March 28th, the water pumps in the secondary cooling system failed and the reactor's turbine broke down. Although the reactor still ran at full power, no electricity was generated, and the high temperature caused the tripping of two devices, one to reduce the pressure, and an eventual "scramming" or shut down of the reactor.

A cycle of disasters followed—the back-up systems that should have pumped water into the secondary system failed because it was closed for maintenance. As the steam generators ran dry, the system lost pressure and an emergency system pumped thousands of gallons of water into the primary system.

Operator inexperience came to the fore, and instead of letting the emergency system's water cool the core, the operator turned off one of the emergency pumps. Unit 2's primary coolant level dropped too low and the water flashed to steam. As the temperatures inside the partially exposed reactor core soared, the fuel rods ruptured and radioactive gas escaped.

Further inexperience caused more disaster. The public officials and utility people didn't move quickly enough resulting in more radioactive exposure to the people. Pennsylvania's Governor Richard Thornburgh took two days to decide to evacuate 3,500 children and pregnant women within five miles of the plant. The majority of 200,000 residents around the plant had already fled.

The Metropolitan Edison, workers of the TMI plant, totally lost the records of emission during the first two days of the disaster, either intentionally or otherwise. Afterwards, residents showed symptoms of high level radiation: metallic taste, hair loss, open sores and high white blood cell counts. Dr. Ernest Sternglass, Professor Emeritus of Radiological Physics at the University of Pittsburgh's School of Medicine arrived on the scene the day after the meltdown. He carried a Geiger counter which definitely showed elevated levels of radiation.

Later, studies done by Dr. Gordon MacLeod of the Pennsylvania Department of Health found abnormal numbers of babies born with serious thyroid problems, and he found a dramatic increase in mortality. Before he could warn the public, he was fired. Later studies were done by Marjorie Aamodt, a former psychologist, and her husband and were carried out around the neighborhood. Cancer death rates around the plant were several times the national average.

The broad lesson of Three Mile Island is frightening for two-thirds of Americans who live in states with operating nuclear plants. Radioactive releases from all plants have lethal effects, and chronic and internal exposure to radioactivity from nuclear plants may be harmful.

In many cases, governments seem to be concealing the radiation effects from concerned citizens. Radiation problems are being worked on extensively by environmental groups, and a number of citizen action networks are built around nuclear reactor plants and other networks around nuclear weapon testing. Looking at the larger picture, nuclear reactors are associated with reducing problems of the greenhouse effect. Nuclear plants built as replacements for fossil fuel plants could conceivably reduce the carbon dioxide emissions by upward of 28% if all fossil fuel plants were replaced in the U.S. by nuclear facilities.

Bill Keepin of the Rocky Mountain Institute estimates that a replacement of coal-burning power plants to nuclear capacity would mean an increase at the rate of one large nuclear plant every 1.6 days for the next 38 years. However, nuclear plant construction costs have escalated exceedingly making their replacements not economical, and a more dangerous problem is the lack of safety of nuclear reactors. No new reactors have been ordered in the United States since 1978, and all ordered since the years 1974-1978 were canceled. Safety factors in question are premature aging of reactor vessels, containment systems that may fail in a major accident, and 30,000 mishaps in plants during the decade of 1979-1989 including a partial meltdown at Three Mile Island.

Environmentalists decry a need for nuclear energy, and believe carbon dioxide emissions can be lowered, and energy needs met by using non-polluting renewable resources, such as solar power, wind and geothermal energy. Christopher Flavin, vice president of Worldwatch Institute, says, "If it were either use nuclear or destroy the climate...I would have to reconsider. But I believe there is a whole range of other options that have been pursued only in a very half-hearted way."

Nuclear power can eliminate only a small portion of gases causing global warming—specifically those from fossil-fueled electrical generating plants. These plants are responsible for less than 17% of the U.S. emissions contributing to global warming. Motor vehicle fumes, deforestation, and industrial and agricultural activities account for the rest of the problem—83% in the U.S. and 96% worldwide.

In order to understand radiation problems, a brief description of radiation seems in order. Radiation consists of particles or waves emitted by unstable atoms, and is a spontaneous decay or disintegration of an unstable atom. If neutrons are added or taken away from the nucleus of a stable atom, the energy balance becomes uneven and something must be expelled to gain balance.

Different radiation types include alpha, beta, gamma radiation, X-rays and neutrons. An alpha radioactive nucleus decays by emitting an alpha particle of two neutrons and two protons. The element then changes to a different element with an atomic number less by two. In beta radiation, a neutron converts to a proton or vice versa, and the beta particle is expelled to balance electricity and release energy. When alpha and beta particles are emitted, the nucleus becomes unstable and excess energy is released as gamma waves (electromagnetic radiation). Neutron emission occurs when the nuclei of heavy, unstable atoms go through spontaneous fission, break into two large pieces and release several free neutrons.

NATURAL AND MAN-MADE RADIATION

We are exposed constantly to natural radiation from cosmic rays and terrestrial sources such as radioactive rocks and minerals in the Earth's crust. Cosmic rays consist of atomic nuclei, mostly protons and can include galactic cosmic rays, radiation belts surrounding the planet and burst of radiation from the sun. Galactic cosmic rays come from stars in the universe, and radiation belts surrounding the planet are called Van Allen belts. They follow the Earth's magnetic field with many charged particles. The sun emits bursts of radiation or solar particle events with sunspot activity, solar flames and geysers of flames. As the flames subside, high energy particles and X-rays spread out in waves.

Terrestrial radiation comes from the earth and is emitted from the surface crust. Certain elements decay, and give off charged particles or gamma radiation. Some radioactive elements are Potassium-40, Uranium-238, Thorium-232 and Rubidinum-87. Some areas of the earth are known for their high radioactivity, such as southwestern India where the sandy beaches contain one-third thorium. Other radioactive areas include Brazil near Rio de Janeiro, France, Egypt, and small islands of the Pacific.

A subsequent problem from terrestrial radiation is food and water coming from regions high in radioactivity. For example, Brazil nuts grown in Brazil are radioactive due to the high gamma rays from the soil, but the nuts don't threaten health. The mineral's radioactive potassium and calcium do pose some threats to health, since potassium contributes up to 1% of the human body, and calcium chemically resembles radium and can accumulate in the bones. Strontium-90 from fallout of nuclear weapon testing can contaminate calcium products.

The danger from man-made activity started when man began splitting atoms to build bombs. At this point, huge quantities of radioactive by-products were generated. Atomic bombs and nuclear reactors produce radioactive alpha and beta particles that are deadly in minute quantities if inhaled or swallowed. Contributing factors in generating larger quantities of radioactivity are weather conditions and the ecological food chain.

Alpha radiation can rip into the cells of internal tissue and create serious damage. It is given off by plutonium, a by-product of nuclear fission, by radon which enters the environment by uranium mining and milling, and radon gas which is a "daughter" element of radium often carried into our homes through the natural environment. Beta radiation consists of a stream of electrons traveling close to the speed of light. The neutrons in the nucleus break down into protons and electrons—the proton remains in the atom and the electron shoots out. Beta rays can cause skin cancer if inhaled and swallowed, and are emitted from nuclear bombs and power plants. Beta radiating products include Strontium-90 and tritium. Gamma radiation is an electromagnetic or wave energy like X-rays, radiowaves and light. It's emitted from a nucleus when it goes through transformation, is highly energetic, and can only be stopped by an inch of lead or iron, eight inches of heavy cement or three feet of sod.

Types of radiation that harm us most today come from radiopharmaceuticals used by the medical industry, X-ray treatment, fallout from atmospheric nuclear weapons tests discharges from nuclear power plants and radioactive waste. Probably the most harmful can be low-level ionizing radiation. We are constantly exposed to planned and unplanned releases of radioactive elements which decay at different rates and enter the food chain at different levels. Radiation can attack the body in three ways; outside bombardment, ingestion with food or water, and inhalation into the lungs.

RADIATION HISTORY

This is only a partial history of events of importance in the nuclear energy movement and nuclear weapons testing. A larger history is beyond the scope of this book. The Atoms for Peace program was launched by Eisenhower at the United Nations in 1953. It was started to promote global sales of nuclear power plants, and to control global supplies of nuclear fuel. By 1954, Congress tried to collaborate with the federal atomic energy establishment by offering partnership with the U.S. Atomic Energy Commission (AEC). The U.S. government next promoted the benefits of atomic energy by funding a Power Reactor Demonstration Program.

The Price-Anderson Act of 1957 was passed to provide for a limit of $560 million to be paid as compensation for atomic accidents. Later in 1987 it was increased to $7 billion. The date is unsure, either December 1957, or January 1958 when an accident in the Central Ural Mountains of Russia contaminated thousands of square miles. The source of the accident was Cheylabinsk-40, a plutonium producing facility erected to produce plutonium to carry out the first Soviet bomb test in August 1949.

The next accident involved plutonium production at Sellafield, Great Britain on October 10, 1957. This event released up to 150 times the maximum permissible level of radioactivity. Local milk supplies became contaminated with radioactive Iodine-131 affecting the thyroid glands. 2 million liters of milk from cows in the immediate area was poured into seas and rivers.

After much use of above the ground explosions, the Limited Test Ban Treaty was enacted in 1963 and restricted testing to underground explosions and obligated the U.S. and USSR to produce a Comprehensive Test Ban Treaty. The Nuclear Non-Proliferation Treaty (NPT) was established in 1970. Under this treaty, non-nuclear-weapon states agreed not to acquire nuclear weapons, and to place all peaceful nuclear activities under safeguards. The treaty was rejected by France, China, Israel, South Africa, India, Pakistan, Argentina and Brazil.

An AEC study later found to be flawed was the Rasmussen Reactor Safety Study (RSS). The Bulletin of Atomic Scientists later warned that the RSS was an in-house study by an agency anxious to get criticism off their back. The EPA published a critique of the RSS and found it had underestimated cancer risks from reactor accidents by a factor of between 2 and 10.

The Threshold Test Ban Treaty of 1974 limited nuclear test explosions to 150 kilotons. This treaty is honored by both the U.S. and USSR. The Nuclear Regulatory Commission replaced the AEC in 1975. In the decade of the 70s, Three Mile Island had a serious nuclear accident with a partial meltdown.

The 1980s continued America's love affair with nuclear energy weapons. The Strategic Defense Initiative (SDI) or "star wars" antimissile program was begun by President Reagan in 1983 and was an ambitious effort to develop weapons to intercept and destroy ballistic missiles in flight. The disastrous Chernobyl meltdown occurred in Russia in 1986. The Intermediate Range Nuclear Force Treaty (INF) was signed in December 1987. Finally in 1988, the Department of Energy shut down the last operating tritium-producing reactor at Savanna River plant in South Carolina. Tritium is an isotope of hydrogen used in making nuclear weapons.

Currently, 128 nations have signed the Limited Test Ban Treaty which prohibits tests in the atmosphere, outer space, or in the ocean. A few hundred strategic weapons can destroy totally the structure of any country. There are now 30,000 such weapons ready for instant use by the U.S., Soviet Union, and France. France is still detonating lethal nuclear weapons. Large nuclear reactor accidents have occurred at Savannah River (1970), Millstone (1975), Peach Bottom and Three Mile Island (1979) and Pilgrim (1982).

One of the oldest government plants producing plutonium is located at Hanford, Washington. Dr. Alice Stewart, a British medical doctor and epidemiologist, conducted studies in the mid-1979s with Dr. Thomas F. Mancuso, an epidemiologist from the University of Pittsburgh. Mancuso had contracted to study worker health at the Hanford plant and found rare cancers 10 to 20 times higher than normal in Hanford workers who were healthy men in their prime of life. A furious nuclear industry and government cancelled all of Dr. Mancuso's funding and confiscated his records. Luckily, Mancuso kept duplicates of his records for later study.

The original 1970s studies were recently verified by an 18 member panel of scientists and experts, who also found alarming incidences of cancer among people who lived down wind from the Hanford Nuclear Reservation. In the years of 1944-1947 residents were exposed to high levels of radioactive iodine. Some infants received doses as high as 2900 rads over three years. One rad is the amount of radiation body organs absorb from a dozen chest x-rays. Iodone-131 if concentrated in the thyroid gland can cause cancer and related diseases.

The radioactive iodine released from chemically dissolving spent reactor fuel rods to produce weapon grade plutonium and uranium was responsible for all the thyroid gland problems.

CELL AND GENETIC EFFECTS OF RADIATION

The genetic effects of radiation are well documented from research studies. The number of genes in humans has been calculated at about 20,000 genes, and these are units of heredity located on chromosomes in the nuclei of cells. All the genetic material is contained in the DNA molecule, but altering the DNA is like shooting a bullet into a computer. When the data carried by DNA is disturbed, it is called a mutation—it can be induced by chemicals or radiation. Radiation can affect any part of the body through the deep penetration of gamma rays or X-rays. Mutations are most serious in the sex cells, and basic alterations can cause many defects to resulting offspring. These defects can include metabolic changes affecting food pathways, an increasing tendency to develop disease, lower intelligence and many other defects.

Radiation attacks the body at the cell structure and may have several effects. It can pass through the cell without doing damage, it may damage the cell, or it may kill the cell and may damage the cell so much that the damage repeats itself when the cell divides. Dr. Alice Stewart compares a radiation damaged cell to a broken plate. The plate can be glued together, but the original integrity is not the same. Every time the cell is stressed it is more prone to break. The repaired cell many not react to disease or injury like an undamaged cell.

HOW CANCER IS INDUCED

As long ago as 1956, a committee on the genetic effects of radiation from the National Academy of Science in "Science" magazine said, "concern about the possible genetic effects due to the fallout on grazing or cropland; fallout in the sea and possible concentrating in marine organisms; the distribution of fallout material by the winds and in the upper atmosphere, possible pathological damage due to long lived isotopes built into our bones. Even small amounts of radiation unquestionably have the power to injure hereditary material."

Dr. Ernest Sternglass, Professor Emeritus of Radiological Physics at the University of Pittsburgh remarks, "We were mislead by experiments showing it took thousands of rads—you get 1/10 of a rad a year from normal background radiation—to rupture a cell membrane. Attention only focused on large levels of radiation and damage to genetic material. There is a completely different biological mechanism that causes damage. Radiation doesn't act on genes, but produces highly reactive chemical species called free radicals." Free radicals can be unstable ordinary elements like oxygen. With a negatively charged electron attached to the oxygen molecule, when radiation enters the cell, the molecule goes through the cell membrane, unzips the cell and causes death. Radiation when ingested or inhaled can sit in the bones and irradiate bone marrow and irreparably damage the immune system.

All nuclear power plants release radiation periodically into the air and water. Such radiation causes cancer, leukemia, birth defects and chronic illnesses in humans. The amount of radiation to cause these illnesses varies in different people. There is no "safe" level of ionizing radiation exposure, and only recently the connection between exposure to low-level radiation and health has been made public. It may take as long as 15-30 years after radiation exposure for cancer to develop.

Dr. Irvin Bross, director of biostatistics at Roswell Park Memorial Institute, said that a federal study of seven counties closest to a reactor shows a "Drastic increase in leukemia rates" in the area. Leukemia arises 2-15 years after exposure to radiation.

Jay M. Gould, principle investigator of a Radiation and Public Health Research Project and member of EPA's Advisory Board from 1977 to 1980 and Ernest J. Sternglass, Ph.D. of the University of Pittsburgh, published a paper in "Chemtech" in January, 1989. They relate that the Chernobyl accident was the largest disaster created by humans. In a few days, it released into the biosphere nuclear fission products equal to a tenth of the amount released by all bomb tests since 1945.

This low-level radiation from Chernobyl arrived in the U.S. about May 9, 1986. An increase in mortality followed with the deaths of 20,000 to 40,000 Americans accelerated during the four summer months of 1986. The evidence was drawn from a number of disciplines: biochemistry, medicine, radiation physics, statistics, epidemiology and ornithology.

In Europe, radiation levels increased to 100-1,000 times higher than the U.S., but the summer increase in mortality increased only by a factor of 10. Low-level radiation follows a logarithmic response pattern, and this response follows the current facts. Europeans were warned not to drink milk, and this probably reduced fetal exposure.

John W. Gofman, M.D., Ph.D., a leading scientist and Chairman of the Committee for Nuclear Responsibility published this letter. "The cancer risk from low doses of ionizing radiation is far more serious than admitted by any of the radiation committees...what's more...this information was available to the radiation community long ago. It is important to remember that all aspects of the effort to prevent nuclear pollution of this planet...hinge on correcting the 'safe dose' falsehood and on correcting the SEVERE under estimates of radiation risk asserted by the radiation community.

"If we permit falsification of the cancer risk from radiation the result is guaranteed to be even more carelessness than is now occurring in operation of nuclear power plants, their fuel cycle, and every aspect of nuclear pollution. The nuclear industry might even succeed in doing away with evacuation plans completely.

"And if the radiation community successfully sells the false idea that even a major accident like Chernobyl will cause no serious health effects, it follows that there is no needs to be bothered with evacuation plans, better monitoring networks, better training of nuclear reactor operators, better earthquake reinforcement, better containment structures and practices at civilian and military facilities, better protection for nuclear whistle-blowers, better maintenance, better cooling ponds, better waste transport, better stabilization of uranium tailings, better precaution in burning an burial of low-level wastes, and in final placement of the high-level wastes...."

At Sellafield, England, the British Nuclear Fuels dumps one million gallons of nuclear wastes containing plutonium into the Irish Sea every day. The government-owned complex is enlarging its capacity for nuclear reprocessing and is getting spent fuel from

Japan, Spain, Netherlands, Belgium, Sweden, Switzerland and West Germany.

The May/June Center Interstate Rad-Waste Report of Kansas City, Missouri says, "For several years, childhood leukemia in the area has been known to be ten times the national average." In a February 17th article called the Gardner report, "The British Medical Journal" rocked the nations by tentatively tracing the cancer pathway to paternal semen. The fathers were typically exposed to cumulative externals doses of no more than 10-30 rem over seven years...On April 5th on Thames TV, Dr. Jill Birch of the Cancer Research Campaign said, "If this Sellafield leukemia is due to germ cell damage sustained by the father, then that means that the child has a genetic form of leukemia, and would be capable of passing that defective gene onto its children."

State Representative Maria Holt and Elizabeth King are Citizen Monitors of the Maine Yankee plant and summarize their efforts in the book, *Monitoring Maine Yankee*. "After the partial meltdown at the nuclear reactor at Three Mile Island, a group of concerned people in midcoast Maine decided to design an early warning system for themselves in case the local nuclear power plant, Maine Yankee, were to experience a similar accident. The group learned that the major release of radioactivity from Maine Yankee is in the form of radioactive Xenon-133 gas, which, though chemically alert, sends out gamma radiation which acts on the body in a manner similar to X-rays. Twenty-five instruments were made that continuously detect gamma radiation. They are present to alarm when the ambient gamma radiation reaches ten times the normal background radiation. Members of the group keep careful records of weekly background readings, and were surprised to experience a number of alarms. These events correlate well with recorded releases from the plant and local weather conditions at the time."

They found that Zenon-133 gas also gives off beta radiation, which doesn't travel far or penetrate deeply into body tissues, but if the gas enters the body by inhalation, it can damage tissue by beta radiation of adjacent cells as well as by more deeply penetrating but less destructive gamma radiation."

The Citizen's Monitoring Network included a wide spectrum of people including a professor of Physics, a radiologist, an electrons professional, a high school Physics teacher, a public health nurse, someone from the army who worked with Health Physics, and others with a technical background. They found a need for independent monitoring 24 hours a day, and asked themselves, if we have smoke detectors at home, why not a radiation detector to monitor a potential accident at the Maine Yankee nuclear plant.

From monitoring over a period of ten years, longer than any other nuclear power plant monitoring in the U.S., they found episodes of alarms on their monitors which correlated with records of radioactive releases from the nuclear plant. They work with several types of monitors—twelve pole monitors surrounding the plant feed information to computers in the office of a State inspector who also accesses via computer to safety parameters monitored in the control room of the plant. The information will soon go to the Health Engineering Department in Augusta. The drawback is that this system is only monitored 40 hours a week.

With this problem in mind, fifty real-time monitoring instruments are located outside the plant in people's homes. They're set to alarm when incoming gamma radiation exceeds a pre-set level. The units connect to a clock which stops if no one is at home, and the exact time of the alarm is recorded. Readings are routinely taken once a week and sent in monthly to Health Engineering to make sure the instruments are working.

Maria Holt remarks, "Radiation leaks are always serious. Maine Yankee had a failed seal problem in November 1989 in which 1,000 to 3,600 gallons of contaminated water spilled inside the plant necessitating the venting of gases. The Nuclear Regulatory Commission allows ridiculously high levels of releasees, so it is very easy for plant personnel to say 'we never release more than 10% of N.R.C. limits'. Whenever we permit releases of radioactivity—we permit harm."

The Citizen's Monitoring Network is also concerned about a nuclear waste dump to be located in Maine in the future. Radioactive material could escape from a nuclear waste dump. Maria Holt says, "The Maine Low Level Radioactive Waste Authority is going through a sitting process, first by eliminating areas unsuitable for nuclear waste (wetlands, natural resources protection areas, aquifers). The Authority is an autonomous body appointed by the Governor. The state law will require legislative approval of a site, as well as a statewide vote and a 2/3 vote of the community chosen to 'host' a site. This will prove difficult."

Maine Yankee produces all of the commercial high-level waste in Maine. This waste must be isolated from the environment for thousands of years. A problem surfaces in disposal of spent fuel rods. Maria Holt says, "A state law passed in 1982 prohibits any spent fuel rods older than three years to be stored in the spent fuel pool after 1992. Since the federal government has no place to store spent futile, and will not likely have a place by 1992, there will be a conflict. The federal government did not intend for spent fuel rods to be stored at reactor sites, much less in pools housed in ordinary buildings, when it planned the light reactor program. The federal government has responsibility for 'high level' nuclear waste."

Is any area around a nuclear reactor plant safe? Maria Holt answers this question, "I do not consider any area around a nuclear plant as 'safe.' Long-lived and dangerous radioactive isotopes are emitted from nuclear reactors routinely. It is a matter of chance who is affected by them. In an accident scenario, large amounts of these isotopes are released into the environment. I can say that I'd prefer to live outside a 50-mile radius of any nuclear reactor to make it less likely to be contaminated by either routine or accidental emissions."

An accident could cause deaths as far away as 225 miles from the plant and dump radioactive wastes on the land and water, subsequently destroying the fishing and agriculture business.

Another similar story emerges from the New York area with an activist couple. Larry Bogart with the assistance of his wife started a Woman's International Coalition to Stop Making Radioactive Waste. They both work from their home in New Jersey. His credentials are impressive—he originally was an assistant to the Chairman of the Board of Allied Chemicals from 1953-1963, and at first was excited about the peacetime use of atomic energy. Later, he became enlightened about the perils of atomic energy use when he did environmental consulting. In recent years he became the Executive Director of the Citizen's Energy Council.

In earlier years at Allied Chemical, he found the biggest danger was a lack of ability to handle radioactive wastes. Traditionally, an engineer follows a process through all the way and that wasn't done with nuclear power. The incidental manufacture of plutonium, such as the modern production of a thousand megawatts in a 10-month fuel cycle generates 500 pounds of plutonium in fuel rods. A neutron is aimed at the nucleus which turns it into a 1,000 fragments. Strontium-90 and Cesium-137 are fission products, and when the uranium is added, transuranic activation products are generated which are lethal. We have created a host of substances that imitate bodily elements. Tritium is poisonous and can preferentially bombard embryonic tissues with an effect that can affect the organism all its life. It has produced an epidemic of brain cancer in children. Most of the pressurized water reactors use boron which creates large amounts of tritium.

Bogart has given over 2,000 talks and traveled a million miles in an effort to organize scores of study groups. He supervised two national opponent networks, and helped prevent 80 proposed nuclear plants from being built. He believes the U.S. has a large nuclear capacity, and that we are vulnerable to a nuclear catastrophe worse than Chernobyl.

Their newsletter, "Women's International Coalition to Stop Making Radioactive Waste", tells the story of the Indian Point nuclear reactor located 25 miles from New York City in the tri-state area of New Jersey, New York and Connecticut. The Indian Point area lies on land bordering the Hudson River, and was once occupied by the Kitchawoke Indians years ago.

Over 300,000 people live within 10 miles of Indian Point and nearly 20 million live within 50 miles. The art and financial capital of the world and the United Nations are within New York City, 25 miles south of Indian Point. After the Three Mile Island accident, Charles Luce, Chairman of Con Edison admitted that Indian Point was built too close to New York City.

Indian Point's nuclear plant has 3 reactors and generates much high level waste. Unit 1 reactor was designed by Babcock and Wilcox (the same firm that designed the ill-fated Three Mile reactor). This reactor had many problems in basic design and was shut down in 1974 by the NRC because it didn't meet new safety regulations. During its 12-year lifetime, it produced power less than half the time it was operated and produced waste that will be toxic for centuries.

The Unit II Reactor and the Unit III Reactor are twins coupled to a larger generator and have more safety features. Justifiable concern is evidenced in power reactor problems—the four steam generators are prone to cracks. The worst one had 42 cracks, and the worst reported crack was 2.25 inches long and 0.53 inches wide.

Each year at Indian Point, there are approximately seven SCRAMS—emergencies that required immediate placement of control rods into the reactor vessel to "quiet" the out-of-control heat and energy inside. The rods are flushed through pipes which are now old and worn—the whole nuclear assembly is aged and worn out. Bogart says, "New York press hasn't warned the public enough about the plant. Years ago, the N.Y. Times gave plenty of space to this issue. It's lunatic to locate a nuclear plant close to 1/10 of the population of the U.S. If a severe or catastrophic accident happened, there would be $300 billion in property damage. No business or homeowner could get damages. There is a $7 billion limited damage claimed by the Price-Anderson Act. The inspector resigned in protest when Con Edison refused to make changes in the plant. Over 150,000 school children couldn't get out during school hours. Trying to arrange an evacuation is as futile as rearranging the deck chairs on the Titanic. Con Edison cancelled two more plants to go next to Indian Point right across the street."

At Chernobyl a 50-mile radius was initially evacuated and at Three Mile Island, radiation was released in 5 minutes. Three years later, another 200 miles were evacuated at Chernobyl and now they are considering evacuating another 200 miles.

Robert Pollard, Union of Concerned Scientists and former safety inspector for the NRC calls Indian Point "an accident waiting to happen." The utilities admit a serious accident is possible and Con Ed's Unit 11 consistently receives low safety ratings.

Bogard is proud of the Woman Coalition, and its second birthday is going to be celebrated, shortly. "It was started when a Women's Writer's Group at Skidmore invited me to take about nuclear problems. At the scene of all nuclear reactor victories, the women started the revolt. Governor Rockefeller wanted a Fast Breeder near the capitol, but the women protested next weekend and got a fair hearing to prevent the Fast Breeder from being built. Women organized a balloon release at Easton on the Hudson River to show that radioactive gases vented over Vermont. The balloons had tags and asked the finders to return the tag. A "Life" photographer on vacation took pictures of the balloon releases."

"A Midland, Michigan woman felt it necessary to oppose plants when she learned of their hazards. We went to Michigan and started the Saginaw Valley Reactor Group. She stopped the nuclear plant from being built in Michigan. Women have stopped more than 80 nuclear plants from being built—there is a similar pattern in Europe."

NUCLEAR WEAPONS TESTS

During the period of 1951-1958, the U.S. did more than 100 atmospheric tests at the Nevada test site located 120 km northwest of Las Vegas. A widespread radioactive contamination took place, and this landed on the downwind area over ranchland and small towns of Nevada, Utah and Arizona. Some of the worst atomic tests were operation Plumbob in 1957 resulting in radioactive exposure to many military troops, and Shot Harry, taking place in 1953. Shot Harry carried an immense cloud of fallout to downwind areas.

"We have grasped the mystery of the atom and rejected the Sermon on the Mount. Ours is a world of Nuclear Giants and Ethical Infants." General Omar Bradley. (From an Associated Press story in the Las Vegas Review Journal, March 11, 1990.)

"A powerful nuclear weapons test at the Nevada Test Site rocked the desert Saturday, just hours after sources reported the Russians would phase out testing in the Soviet Union mainland because of safety concerns."

The first 1990 nuclear test was an underground shot, code-named Metropolis. It lies buried at the bottom of a 1500 foot shaft at Yucca Flats which is 85 miles northwest of Las Vegas. Shock waves generated by the test measured 5.1 on the Richter Scale in Golden, Colorado, and had an explosive force of 20,000 to 150,000 tons of TNT. Right after the 8:00 a.m. detonation, eleven anti-nuclear protesters turned out at the gates.

The protesters at the Nevada Test Site may have come from the Nevada Desert Experience, a faith-based organization with Franciscan origin, and scriptural values working to end nuclear weapons testing through a campaign of prayer, dialogue and non-violent direct action. It was first organized in 1984 and now offers both a August Desert Witness to commemorate Hiroshima and Nagasaki and a Lenten Desert Experience.

The Nevada Test Site is 65 miles northwest of Las Vegas in Nye Country, and is operated by the U.S. Department of Energy to provide a site for testing nuclear weapons. Every three weeks, new tests are conducted at a cost of 6-70 million dollars per test. Historically, on January 27, 1951, the first atmospheric test was conducted on the 1,350 mile area.

The Nevada Test Experience (NDE) actions occur during the Lent season in the spring, and the August Desert Witness marks the anniversary of the Hiroshima-Nagasaki bombings that ended World War II. Since 1984, NDE has held more than 20 test site actions involving over 10,000 people, and has organized test ban vigils in Las Vegas and other communities, has distributed material about testing and nonviolence, held several conference, initiated a newsletter on nonviolence and weapons testing and has worked with various religious communities to end testing at the test site.

CITIZEN ACTION

The Public Citizen Critical Mass Energy project suggests two strategies for closing nuclear plants; ballot measures to prohibit new plants and close existing ones, and efforts to establish publicly owned utilities. If citizen groups use ballot measures to close nuclear plants, they must develop a strong base of support for the campaign and must qualify to get on the ballot by collecting signatures from registered voters.

Citizen sponsored ballot measures face several obstacles including the large amounts of money the nuclear industry has for campaigns, if a proposal fails they may claim the public supports nuclear power, and lengthy and expensive legal challenges that cause court battles. However, nuclear initiatives can provide a direct way to close existing plants.

Many consumer groups support public ownership of electrical utilities. When these are publicly owned, they can offer electric service 30% less than private utilities. At present, sixteen nuclear plants are publicly owned and many are partially owned by municipal utilities and cooperatives. Consumers can authorize either a closing of the nuclear plant or can sell their share of the facility. How can this be done? Citizens can elect people who oppose nuclear power to the utilities board or directors or to the city council.

7 AIRBORNE TOXICS ACTIVISTS

Representative Henry A. Waxman (D) of Los Angeles is a key player in his efforts to pass a tough law eliminating air pollution, acid rain and airborne toxics. Of short stature, with a moustache and almost bald, he's known as Mr. Clean; he believes clean air is not just a Los Angeles problem, but one concerning 150 million people who live in places violating the Clean Air health standards.

Waxman's efforts on clean air legislation are felt in the following ways: tightening tail-pipe standards for hydrocarbons and nitrogen oxides, stringent standards for light-duty trucks, and stricter standards for heavy-duty trucks and urban buses. He feels they should have locked in the pollution requirements of the auto for the full life of the car or 100,000 miles in ten years—not just for five years and 50,000 miles.

Through the efforts of Waxman, complex compromises in the Clean Air Act were won in the spring of 1990. Congress passed the new resolutions by an overwhelming vote including stronger sections on alternatively fueled cars, stationary sources of offshore oil pollution. The Waxman-Lewis amendment established a national clean-fueled car program. All severely polluted cities are required to use ultra-clean fuel buses and fleet vehicles by 1994. California will lead the way with a pilot program offering 1 million clean-fueled cars for sale to consumers over five years beginning in 1994. Representative Jerry Lewis (R-Redlands) said," This is a historic day for the people of Southern California and the country. Once again, California is leading the national clean air debate by implementing a program that will lead to widespread use of clean-fueled vehicles."

The projections on the price tag on the new Clean Air legislation for consumers and industry varies widely. President Bush wants to veto any clean air bill whose cost exceeds by 10% the estimated $20 billion to $21 billion annual price tag put on the Administration's own clean air legislation proposed in 1989. Waxman remarks, "We're convinced that the health and benefits will far outweigh the costs. The costs will be below the official estimates and in the end the Clean Air Act will actually stimulate economic growth rather than impair it."

What are the various government agencies, private citizens, and environmental groups doing to stem the tide of toxic pollutants in the air? One of the most innovative and far reaching governmental plans is implemented in the Los Angeles area through the South Coast Air Quality Management District (AQMD), and their solutions to this problem are covered further along in the chapter.

What are the common air pollutants and where do they come from? You are risking your health daily with pollutants in the air you breathe, and these toxic air problems must be solved with a combination of various efforts. Carbon monoxide has effects on the heart and brain, sulfur dioxide from acid rain can promote lung damage, nitrogen dioxide causes bronchitis and pneumonia, ambient lead leads to brain and nervous system damage, and suspended particulates can cause breathing problems.

Exactly what is smog, and how does it affect your life? Smog stands for a combination of the words, smoke + fog, and is often used for photochemical smog to describe a mixture of oxides of nitrogen, ozone and complex organics created by the action of sunlight on urban pollutants (such as Los Angeles and Tokyo.) Nitrogen oxides (produced by the combustion process) can combine with volatile organic chemicals (gasoline vapor, paint thinner, dry cleaning fluids and other industrial chemicals) in the presence of sunlight to form ozone, an irritating chemical that is the chief component of smog.

The U.S. government has been aware for years of the hazards posed by dirty air. The original Clean Air Act passed in 1970 instructed the EPA to identify toxic air pollutants and regulate them by setting and reinforcing standards that provide, "An ample margin of safety to protect the public health from such hazardous air pollutant." When the Clean Air Act was revised, the EPA established clean air standards for only five toxic air pollutants, although several hundred were identified.

In 1989, President Bush offered the American people a new Clean Air Plan: its features were a 50% slash in acid rain producing sulfur dioxide emission by the year 2000, a 40% tightening of emission standards for hydrocarbons from auto tail pipes, and a 75% cut in cancer-causing toxic chemicals poured into the atmosphere. In conjunction with President Bush, the EPA ranked industries according to toxics released, for example, the chemical industries, smelting, pesticide production, petroleum refineries and tire manufacturers. The EPA would evaluate the cost of clean-up technology, and order toxic production sites to use these technologies. They could order refineries to light flares on smokestacks to burn off organic chemicals, and order smelters to install fabric filters on stacks to trap toxics. These chemicals from plants are dangerous as Lloyd Connelly (D-CA), State Assemblyman comments, "Almost every manufacturer and user of toxic chemicals that has been examined routinely releases significant levels of substances proven or suspected of causing human health hazards including cancer and birth defects."

In order to understand toxic air problems, concerned citizens should learn about three common air pollutants that cause the most harm. Acid rain is briefly surveyed since an entire chapter will be devoted to it later. The other two toxic pollutants include industrial pollutants from all different kinds of industries, and the current problems from smog and ozone in some of the larger cities in the U.S. and the world.

A comprehension of acid rain requires a working definition of the term. Rain and snow can become dilute sulfuric, nitric and hydrochloric acid by interaction between the sun, wind, water and chemical pollutants, mainly sulfur dioxide and nitrogen oxides. These chemicals are released into the air wherever coal, oil and natural gas are burned: from the smokestacks of electric generating plants, metal smelters and industrial boilers, and from exhaust pipes of motor vehicles. They occur from natural sources as well, such as volcanic eruptions, forest fires and bacterial decomposition of organic matter. However, the most troublesome problem is added upon by human activity.

Why is acid rain so serious and a continuing environmental problem?

It can extinguish whole species of fish and aquatic life; it can reduce the productivity of forests and farmlands, disrupt plant photosynthesis and poison drinking water and fish used as food. In addition, it causes corrosion of buildings, cars and art work, and the health effects can cause people who are susceptible to heart and lung disease from contaminated air.

INDUSTRIAL AND VEHICLE TOXICS

Besides the harm caused by acid rain pollutants, another large area of concern lies in industrial plant emissions. U.S. manufacturers released 2.6 billion pounds of toxic pollutants into the air in 1987 according to "USA Today's" analysis of Environmental Protection data. Ted Smith of Silicon Valley Toxics Coalition says, "Companies have simply been using the sky as an open sewer." These chemicals leak into the air through smokestacks, through ventilation systems and open windows and from temporary storage at plant sites including open outdoor tanks and storage ponds. The Clean Air Act proposed by President Bush would require industrial toxic emissions to be cut 75% within 20 years and this law is badly needed.

It's difficult to separate industrial pollutants from those producing smog, but generally there are several different categories of pollutants responsible for air problems. Sulfur dioxide, part of acid rain, is most important as a primary pollutant, and is a result of direct emissions; it comes from fossil fuels containing sulfur as an impurity. Oxides of nitrogen (NO_x) are formed as combustion products of high temperatures, and atmospheric conversion of nitrous oxide to nitrogen dioxide is an important step in photochemical smog. Another major compound is ammonia, and some of the nitric acid formed is removed by acid rain or dry deposition.

Other important pollutants covered here include carbon derivatives, carbon monoxide and carbon dioxide—industry tries to achieve complete combustion of carbon, because carbon monoxide is highly poisonous. Carbon monoxide has a high affinity for the hemoglobin in blood, and cells deprived of oxygen may die. Ozone can be beneficial as a protective layer in the stratosphere, but at ground level as a portion of photochemical smog, it can be very irritating. Hydrocarbons contain both carbon and hydrogen and in chemical structure can be open chained, single or double bonded, or a closed loop of six carbon atoms called a benzene ring. Hydrocarbons can combine in a series of chemical reactions with oxides of nitrogen, and can be catalyzed by the sun to produce photochemical smog, for example PAN (peroxyacetylnitrate).

LOS ANGELES AIR PROBLEMS

For years a consuming problem occurs in the Los Angeles basin. "The New York Times" calls Los Angeles "a colossal city-state covered with a light brown fuzz." The "Los Angeles Times" told the South Coast Air Quality Management District (AQMD) and the Southern California Association of Governments (SCAG) that to achieve clean air they'll have to simultaneously act as "militant regulators, loud cheerleaders, sound planners and above all effective educators."

Los Angeles is a victim of unusual meteorology and spectacular topography. Ozone and carbon monoxide levels reach maximums nearly three times the national standard set to protect public health. Fine particulate matter reaches levels nearly twice the national standard, and the Basin is the only area in the nation failing to meet nitrogen dioxide health standards.

In spite of forty years of local regulatory efforts, these problems still exist because too many pollutants are emitted into an area, and these exceed the carrying capacity. Seasonal effects contribute to the pollution; in summer an inversion layer concentrates pollutants under a lid of hot air, and in the winter inversion levels occur at ground level during the night and early morning hours causing increased concentrations of carbon monoxide and nitrogen oxides. Sea breezes contribute to the effect by moving contaminated air masses containing reactive organic gases (ROG) and oxides of nitrogen (NOx) inland across the Basin.

Where do all the harmful emissions come from? Nearly all human activities cause the problem including large industrial sources. New automobiles contribute far less pollution than they did years ago, but still mobile sources represent 70% of today's emissions. The AQMD has categorized the pollution into four categories according to activity; on road mobile (cars, trucks, buses) other mobile (airplanes, ships, trains, construction equipment, etc.,) residential/commercial/service, and industrial/manufacturing.

The residents have a unique way of coping with the smog. Vents off; window up; radios on. This is a checklist the typical South Coast commuter reviews each day before his commute. Thousands try to foil the air pollution by sealing their cars from the toxic fumes surrounding them. However, the AQMD shows that air toxics entering cars during peak traffic times expose commuters to between two and four times the levels of cancer-causing toxic chemicals found outside. In fact, toxic pollution concentrations are higher in slow-moving traffic of 25 miles per hour, but improve when the traffic moves freely.

Business entrepreneurs have an equally difficult time integrating healthy business with healthy air. Harvey Clements owns Only

Ovals, a firm that manufactures wooden oval art frames in Orange County. His manufacturing process emits smog-forming compounds, and he must comply with AQMD rules. He comments, "I grew up here. I appreciate the air pollution problems...I want to do whatever it takes to make my company safe to work in and around, while maintaining our bottom line."

His growing business now employs 18 people and brings in $1.2 million in total sales. He trys to shoot for the 1994 compliance with rules limiting emissions from wood coating. He recently changed his spray painting equipment to the new high volume, low-pressure system which was approved by AQMD. "We use less gallons of product per day with the new spray system because much more product sticks. With the conventional spray gun, most ended up in fog around us." He, also, uses a new lower VOC thinner solvent for cleaning his equipment, and continues to experiment with new products. He cooperated with the AQMD's ridesharing effort by organizing a carpool program (five cars for 18 employees), and reimburses each carpool for gas money.

HOW AIR POLLUTION AFFECTS YOU

The EPA has done extensive studies of smog and ozone types of air pollution, and its conclusions come from the Clean Air Scientific Advisory Committee of its Science Advisory Board. These studies include experiments on animals, controlled testing of humans in laboratory or field setting, results of accidental human exposure and epidemiological research (comparing pollution levels and associated health statistics among different communities).

The federal and state governments set health standards for pollutants specifying levels beyond which the air is unhealthful. First, second and third stage smog alerts are based on the degree of health risks. The California state standards are tougher than federal because states are permitted to set stronger standards if they believe they are necessary.

Ozone is not directly emitted into the air, but forms when hydrocarbons and nitrogen oxides react in sunlight. These chemicals may come from motor vehicles, petroleum refining and marketing, oil storage tanks, chemical manufacturing, surface coatings, house and consumer products, dry cleaning and printing industries.

Ozone can threaten your lungs by damaging cells and inflaming and swelling the airways. It reduces the infection fighting qualities of the respiratory system. Beware of ozone if you have asthma, emphysema or chronic bronchitis. In the Los Angeles Basin, alone, ozone has damaged the respiratory systems of 1.2 million people. Ozone can hurt healthy people as well as those at risk. Angelenos find breathing becomes more difficult during work and exercise causing irritation and discomfort.

In a New Jersey summer camp, exposure to ozone for five days damaged children's lungs. In a study done on adult men, the EPA studied 10 men exposed to 0.12 ppm ozone for 6.6 hours (including 5 hours of moderate exercise), and the results showed a decrease in lung function. The symptoms of breathing and coughing increased over the six hours of exposure.

Two groups of people especially at risk from ozone are athletes and children. "Athletes may be relatively young, healthy, physically fit non-smokers, but they may be among the most vulnerable to the effects." Those at risk belong in the following categories.

- Heart patients who have less exercise capacity and experience angina attacks sooner.

- Healthy young men suffer decreased oxygen consumption during strenuous exercise.

- Some people have significant reductions in visual perception, manual dexterity, ability to learn and perform complex sensorimotor tasks.

Other components of smog, such as nitrogen oxides, fine particles known as PM_{10} (less than 10 micometers in size) and hydrocarbons can be quite damaging to the health. Even at lower levels, nitrogen dioxide irritates the lungs, causes bronchitis, pneumonia and lowers resistance to respiratory infections like influenza. Nitrogen oxides also contribute to acid deposition in many areas of the country.

Particle pollution causes both short and long-term reduction in lung function, and contributes to chronic respiratory illnesses, cancer and premature death. Fine particles (PM_{10}) cause much harm because they can reach the deepest recesses of the lungs without being captured by the natural cleansing action of the respiratory system. Long term studies indicate that people living in areas with high particle pollution have more respiratory problems and lower levels of lung function than those in cleaner areas.

Some hydrocarbons are toxic; some are not—currently most hydrocarbons don't present a threat to health. They mainly contribute to the formation of ozone, however, one hydrocarbon may be dangerous.

Dr. Henry Gong, Jr., M.D. reported in 1987 in the "Journal of Sports Medicine and Physical Fitness" about ozone levels effects on athletic performance. The study suggests that performances begin to suffer at 0.12 ppm (federal standard) and increases as the concentration tops at 0.20 ppm. Ozone studies on children raised in the South Coast Air Basin show a 10-15 percent loss in lung function compared to children who grow up where air is not polluted, according to Dr. Kaye Kilburn, M.D., Professor of Medicine at University of Southern California.

The EPA in its Review of National Ambient Air Quality for Carbon Monoxide discovered that adverse health effects may be experienced by these sensitive people above levels of 15 ppm (eight hour average). Currently, the levels are set at 9.5 ppm to allow for a margin of safety.

Since Los Angeles has overwhelming air pollution problems, many involving ozone, smog, and carbon monoxide, all eyes focus on their innovative attempts to solve their problems. The South Coast Air Quality Management District of Southern California hopes to achieve federal air quality standards by the year 2007. A three-tiered strategy moves into place to achieve the air quality standards; the first tier uses currently available technology and management practices, the second tier includes on the horizon technologies that can be expected in the near future, and the third tier researches and develops new technological advances.

An innovative solution capable of use by many cities involves a rideshare fair to get employees interested in cutting the number of autos on the freeway. Often, the rideshare fairs include prize give-aways, refreshments, balloons and live music to serve as an incentive. While enjoying themselves, the employees are exposed to information on company sponsored commuting programs.

Besides carpooling, staggered working hours become mandatory by 1994. Most fleet vehicles used by the government and car rental companies will be driven by electric power, or required to use a clean burning fuel such as methanol. The AQMD hopes to tear Southern California motorists, industries and government agencies from gasoline, fuel oil and other highly polluting materials, and promote the use of better environmental means within 15 years. Within 20 years, all vehicles are expected to run on electricity or a clean power source.

Citizens can expect their daily lives to change. In the household charcoal lighters, paints, solvents and hairsprays will be banned. Bias-ply tires responsible for depressing gas mileage and emitting particulates must be eliminated. Places of business will be required to alter or replace equipment to reduce emissions, and businesses affected will be restaurants, bakeries and dry cleaners.

These measures will change Angeleno's daily lives, and concurrently this tremendous effort is being watched by cities like New York and Chicago who face severe air pollution.

Continuing with inspiring governmental and citizen protest stories, and individual activists efforts, the rest of the chapter deals with ways everyone can affect the quality of the air we breathe. The Ensco story is occurring right now in Arizona, individual activists George Smith and Jerry Tinianow work in Texas and Ohio, and Helmut Ziehe addresses the continuing problem of indoor air pollution.

Arizona, a neighbor to California, is confronting a problem originally started by the state government's effort to dispose of hazardous waste. Ensco, a hazardous waste plant proposal near Mobile, Arizona, had its conception ten years ago. On May 7, 1990, images of protesters being dragged from the site appeared on TV screens all over the country. Greenpeace volunteers used their organization and publicity skills to focus on this problem on Earth Day, 1990. There were planned sit-ins at the State Capitol with a bed chained to a sign reading, "Why is Arizona in bed with toxic polluter?" These images promoted Arizona Governor Rose Mofford and the Department of Public Safety to order an investigation and subsequent public meetings on the issue.

The incineration plant originally was scheduled to service only Arizona's hazardous waste, but later turned into a project importing 70% of its waste from other states. Ensco tells the public that a larger facility is necessary to make it profitable for them. Investigations showed a bad track record in Arkansas, and their plant was rated third worst in the pollution in Arkansas in 1989. Additionally, Ensco was fined nine times from 1983 to 1988.

Many hazardous waste disposal sites end in towns with minority, poverty-stricken populations, and Mobile, Arizona, a tiny town 30 miles southwest of Phoenix qualifies in this category. Most of the waste imported will be PCBs (polychlorinated biphenyl compounds) used in industrial electrical transformers. These compounds do not break down, and have been linked to disorders in the brain and liver, cancer and birth defects. PCBs were banned by the EPA in 1979 resulting in tons of PCBs waiting disposal. Ensco plans to use a high temperature waste incineration.

Arizona fears the incinerator will affect the air and water quality as it spews forth several tons of lead, cadmium, particulates, carbon monoxide and volatile organics. Opposing forces say Arizona needs a smaller facility only for Arizona with no state ownership. The liability for spills and accidents should rest on a private company and not on taxpayers. Finally, Governor Symington ended the Ensco toxic waste dump controversy by buying out Ensco for $44 million dollars, and saved the state of Arizona from a long, drawn-out lawsuit that would have cost the state much more.

INDIVIDUAL ACTIVISTS

In the Southwest in the modern city of Houston, Texas, Dr. George Smith, a dentist with a well-developed environmental consciousness, works behind the scenes in the fight against air pollution. He has a long history of activism on Air Quality Subcommittees in the Sierra Club, EPA Sierra Club Air Quality Education Projects, EPA Air Pollution Research Committees, and was a lecturer at the University of Texas School of Public Health on Houston Air Pollution.

He originally moved to Houston, a city called the Golden Buckle in the Sun Belt, to establish his dentistry practice. This interest pushed him into joining the Sierra Club where he became the state air quality chairman for ten years. He found the environment attacked from all sides in Texas, and especially in Houston where developers put in dams, built subdivisions without controls, cut hundreds of acres of timber and polluted the air from petrochemical complexes.

Dr. Smith says, "One day, a woman named Becky Moon talked to our group of Sierra Club leaders about the dangers of vinyl chlorides being made at several plants out in the ship channel. She had us write letters to the Texas Air Control Board which resulted in the companies reducing their emissions, and agreeing to fund a health conference on toxic air pollution. It was my first experience how an individual working with allied groups can influence state policy and cause industry to be more responsible.

"In Houston, we had dozens of plants with dark plumes from their stacks. The state agency tried to control air particulates and this involved city council votes to grant 'variances' for different plants until they put in control equipment. Progress was slow, but after a few years, the black smoke was gone leaving a residue of toxic vapors. Our smog is somewhat different from L.A.'s where they average over 150 days over the standard, while we average only 40-50 days. Houston smog is more a result of industrial sources—330% area sources and vehicle smog only 33% compared with LA smog of mostly vehicle emission.

"We recently had a small victory. One of our committees saw a notice of Compak Computers seeking to triple their emissions of CFCs. We formed a delegation to go meet with Compak to negotiate with them. We discovered a lack of new equipment; they planned to run more shifts using the same degreasers, and had doubled their emissions with a lack of a permit for the doubling. We negotiated a 24% reduction and an assurance they would phase out CFCs in five years.

"A similar experience happened when we worked with a channel petrochemical plant. Because electric rates and gas prices skyrocketed during a fuel shortage, coal was being considered for the plants. This could have an enormous impact on the air if many plants switched to coal. We set up a meeting with the Texas Air Control Board and the industry to discuss the permit and control equipment. The industry admitted to using a coal which would emit sulfur. They dropped their plan.

"We get little credit when we oppose short-sighted projects, but the most spectacular was the Allen's nuclear plant proposed by

HL&P (Houston Light and Power). It would have been close to town with no zoning and no geographic constraints or concern for urban sprawl. After stiff opposition, HL and P saw the light and dropped the project as energy conservation came to the front.

"Environmentalists opposed the super sonic transport because of noise and damage from sonic boom, and damage to the protective layer in the stratosphere. The Concorde SST has been a financial drain to its countries and loses money every time it flys. No one thanked us.

"I've tried to work with the system. I'm just a little uneasy with Earth First! people using media confrontation and extremist tactics. However, these other groups do seem to have a place and can arouse the passions of people. Last year, we supported bills in Texas to mandate use of clean burning natural gas for school buses, state agency vehicles and metro bus fleets. Smoke and smog-forming emissions are dramatically reduced with natural gas and we are all proud that the bill passed."

In the Midwest, a clean air activist attempts to reduce air pollution problems in Ohio. Jerry Tinianow, a lawyer, first became interested in clean air after he joined the Sierra Club in 1981, and he was appointed to the Clean Air Steering Committee of the Sierra Club. In 1978, he prepared testimony to the EPA on the environmental impact of building a steel mill in Conneaut, Ohio. Jerry generally works on broad policy questions, and not on detailed laws and regulations.

Tinianow is extremely involved in lobbying. He says, "By lobbying, it means to personally visit offices, and I urge them to meet with congressmen or staff people. Usually, it is in regard to sponsoring bills, or co-signing letters, or voting on specific bills. I feel I am good at organizing volunteers and promoting. Learning technical background is easy—organizing and motivating volunteers is hard."

The clean air issues in Ohio are complex and Tinianow gives a background of the problem. He remarks, "Ohio is a big coal-producing state, and has coal rich in sulfur which contributes to acid rain. We are the number two auto manufacturer. Akron, Ohio manufactures rubber and tires, and Columbus has two large Honda plants. Toledo manufactures the Willis jeep and Cincinnati had a big auto plant until recently. We have small manufacturers who make car components. The worst air pollution comes from sulfur dioxide and smog— we are the victims of air pollution. We have a large Congressional delegation of 21 Congressmen, and a population of nearly 11 million—higher than Michigan. The power plants in Ohio burn Ohio coal which contributes to acid rain."

Helmut Ziehe of Florida helps the clean air problems within our home environment. The Bau-Biologie movement started in Germany over 20 years ago, and the German word translates to mean buildings and the science of life. The movement realized our living and working environment may be killing us due to toxic air pollution, radiation problems, electromagnetic radiation, and hazardous and toxic materials. The German people recognize biological destructiveness related to the environment, and recently started another ecological movement, The Green Movement.

Helmut Ziehe, the healthful building authority and pioneer, brought the Bau-Biologie movement to the United States in his International Institute for Bau-Biologie in Clearwater, Florida. He remarks, "In Germany the baubiological movement was way ahead of what's here in the U.S. Both the appearance and the experience of being in a baubiological building are different from other buildings. There's a special, soothing look to a finely constructed baubiological building with its gleaming hardware in and out. There are the smells, usually a fantastic touch of citrus from the plant based paints and finishes. Since the electromagnetic fields have been minimized, and there's a nice balance of ions in the air, you feel clearheaded and alert, the solar heating adds comfort and light."

Helmut Ziehe grew up in postwar Germany and received his architectural diplomas, did postgraduate work in London, and worked with architects in Sweden and Germany. He became aware of the "sick building" syndrome, a problem arising from synthetic, air-tight buildings. At that time, he started paying closer attention to his health, but found that his feeling of health evaporated in certain buildings. In 1986, he relocated to the U.S. to spend his time teaching others of Baubiologie which he disseminates through a rigorous correspondence course.

Ziehe describes the sick building syndrome in a detailed way. He says most of the modern buildings in cities and rural areas are made of artificial or man-made building materials, such as plastics, paints based on petrochemicals and insecticides, glues which are full of formaldehyde, particle board and plastic carpets. Heating and cooling systems can be a very severe health problem, and are often the reason for the "sick building syndrome." Houses today are constructed with air conditioning, and people who live or work in these quarters are forced to breathe recirculated air filled with cigarette smoke, bacteria, body odors, smells from dead animals trapped in the ducts, and dust and mold build-up.

Ziehe believes non-toxic materials can prevent "sick building syndrome", and materials before the industrial era were natural and organic. These materials are chemically untreated wood, clay (adobe and rammed earth buildings), clay bricks, cork, coconut fibers, sisal, wool, bees wax, paints based on natural ingredients and limestone.

He says natural walls, floors and ceilings enhance air diffusion. Diffusible building materials in walls, ceilings and floors allow gas exchange to take place, and the air can be permanently regenerated. Harmful or toxic substances with higher concentrations inside the room are constantly being moved towards the lower concentration. A dilution takes place.

The counter flow of gas results in a supply of fresh air. He speaks of a "breathing wall" or a "third skin", as far as a house built with natural materials and methods is concerned. Vapor barriers and plastic surfaces will stop this process. Plastic changes the balanced distribution of negative and positive ions to a situation where there are an abundance of positive ions available. The positive ions attract dust, bacteria, smell particles and decrease the quality of the air.

Ziehe warns against radioactive building materials for they can give us unacceptable levels of radioactivity in our bodies. Radiation emanating from radioactive materials destroys body cells, and if the dose is big enough will result in body death. If we do not check our radiation level in our homes, we might have unacceptable levels of radiation. Over a long period of time, this fact will cause severe health problems.

There is much talk and controversy over radon in the home. If there is sufficient ventilation in a house, there would not be a problem as the concentration remains low. However, energy-efficient homes with sealed windows, doors and walls and often no artificial ventilation allows the concentration of radon to increase dramatically and thus become a hazard.

It is very difficult , if not impossible, to keep radon out of dwellings. One possible way would be to raise the first floor off the ground, and to eliminate cellars and basements which are too expensive, anyways, and just store rubbish. Another way would be to see there is sufficient air exchange. People can install a device to measure the radon level, and indicate when the windows and doors should be open for ventilation.

As to the threshold level of radon, Ziehe says that radioactivity doesn't have a safe level and this translates into NO RADON. EPA recommends 1 pC/liter—when it reaches 4pC/liter it becomes dangerous. In any case, provide for sufficient air ventilation.

A new area of concern regards the effect of electromagnetic fields on health. Little research has been done in this field, and often the data is biased. From what is known, it seems important to avoid electromaqnetic fields. These artificial fields interfere with the body electric system which has a resonance level in the range of 8HZ. Studies show that other frequencies enhance cancer growth dramatically. Until more scientific studies are done, people should stay away from electromagnetic fields.

AUTO PROBLEMS AND POSSIBLE SOLUTIONS

Because of both air pollution and global warming, the public needs to buy cars that are low in emissions, and have a high fuel economy. New technologies can be used to both improve fuel economy and reduce emissions. Chris Calwell of the Natural Resources Defense Council analyzed 781 car models. The analysis showed that the 50 most efficient cars emitted one-third less hydrocarbons than the average car, and only half as much as the 50 least efficient ones.

Many cities restrict the use of driving, for example in Florence, Italy, the middle of the city is now a pedestrian mall during the day. Mexico City and Santiago keep one-fifth of vehicles off the streets on weekdays based on license plates numbers. Most urban areas need a shift away from autos to public transportation, car pooling and bicycle commuting. All of these methods offer cheap, sensible and effective ways to eliminate air pollution.

Alternative fuels of electricity and hydrogen are most promising. Electricity-run cars with electricity produced by natural gas can reduce hydrocarbon emissions by 99%, carbon monoxide by 100% and nitrogen oxides by 84%. If cars were run on hydrogen produced by electrolysis of of water under pollution-free power sources, there would be no hydrocarbon or carbon monoxide emitted and less nitrogen oxides.

Change in both auto design and types of fuels can be effective. Car engines can be modified and catalytic converters installed to reduce harmful emissions. Converters can reduce hydrocarbon emissions by 87%, carbon monoxide by 85%, and nitrogen oxides by 62% during a vehicle's life. Vehicles can use alternative fuels such as methanol, ethanol, natural gas, electricity and hydrogen. A problem with methanol could be the production of two to five times more formaldehyde, and the possibility of coal-made methanol which could increase carbon dioxide emissions.

Although natural gas could reduce emissions 40-60% for hydrocarbons and 50-95% for carbon monoxide, there is a problem of increased nitrogen oxides. Another alcohol fuel, ethanol, could emit acetaldehyde that would speed ozone formation and cause more ozone problems.

The next chapter goes into greater detail about the acid rain problem. A neighboring country is at the fore front in reducing acid rain problems, and would like to work with the U.S. on this problem more cooperatively.

8 RAIN FROM THE SKY

"We acknowledge responsibility for some of the acid rain that falls in the United States, and by the time our program reaches projected targets, our export of acid rain to the United States will have been cut by an amount in excess of 50%. We ask nothing more than this, in return from You." Prime Minister Mulroney—Address to Congress April 17, 1988.

This quote emphasizes the acid rain problem between the U.S. and Canada. Both countries are obliged by international law to reduce air pollution to an amount that doesn't cause damage to either one. Canada and the U.S. have a long history of mutually beneficial cooperation to protect their common environments. The Canadians are very concerned about acid rain because it's endangering vital economic resources of fisheries, tourism, agriculture and forestry, and currently acid rain is causing $1 billion worth of damage in Canada every year.

More than half of the acid deposition in Canada originates from emissions in the United States. A 1979 study showed both Canada and the U.S. were to blame with the winds from the Midwest contributing to the problem. Comparing Canada with The U.S. shows Canada has fewer cars and coal-fired power plants, but does have gigantic metal smelting industry and processes ores with copper and nickel that also contribute sulfur.

The acid rain emanates mainly from coal-fired power generating stations (major source in the U.S.) and non-ferrous ore smelters (major source in Canada). Most of the NO_x emissions come from vehicles and fuel combustion. Because it's produced in areas of heavy industry and dense population, most of the producers are located in the American Midwest and the Canadian provinces of Ontario and Quebec.

Canada and the U.S. are responsible for the common acid rain and both are making attempts to remedy the problem. In February 1985, Canada launched a major acid rain control program with the federal government and seven provinces east of Saskatchewan agreeing to reduce the total Canadian sulfur dioxide emissions by 2.3 million tons per year. Each agreed to enforce these limits and to achieve them by 1994. Canada will reduce transboundary flow of sulfur dioxide to the U.S. by 50%, and hopes the U.S. will reduce its export of sulfur dioxide to Canada to two million tons, or 50% of the 1980 level.

The American Congress and Senate recently passed Clean Air legislation in the spring of 1990. The provisions of this legislation calls for a reduction in sulfur dioxide emissions from coal-burning utility plants by ten million tons by the year 2000, thus cutting emissions in half. Utilities are allowed to buy and sell "pollution credits" among themselves with extra allowances given to Midwest utilities, and some "clean plants" to ease the cost burden and allow for future growth.

Nitrogen oxide releases must be cut by 2.7 million tons. The sulfur dioxide emission cut will affect 107 coal-fired electric power plants in the Midwest still using high sulfur coal. They face the greatest pollution reduction burden with corresponding increases in electricity rates.

The Midwest utilities and high sulfur coal miners will be affected adversely. The Senate rejected the proposals to ease the cost burden on Midwest utilities, making electrical cost increase for consumers a sure thing. Should the bill include financial help to coal miners who lose jobs because utilities shift from using high sulfur coal? That remains to be adjusted in the future.

Americans could well look at the progress made by four large polluters in Ontario Province. Ontario will achieve a 60% reduction in sulfur dioxide emissions by 1994. Inco lowered emissions by 40% achieving a sulfur dioxide containment rate of

70%; Falconbridge achieved a containment rate of 85%; Algoma Steel has reduced emissions below the 1994 ceiling, and Ontario Hydro has reduced emissions by approximately 25%.

An important part of the cooperation between the U.S. and Canada lies in protecting the Great Lakes from acid rain. Canada and the U.S. take pride in sharing the world's largest undefended border, but this results in transboundary environmental problems. The Great Lakes pollution issue led to the Great Lakes Water Quality Agreement of 1972 to establish objectives, programs and measures to restore and enhance the quality of Great Lakes water.

By the mid 1970s, the Great Lakes came under attack from toxic chemical buildup. By 1978, a revised Great Lakes Water Quality Agreement was signed with the governments having a commitment to protect the ecosystem of the Great Lakes basin. This agreement adopted the policies of virtual elimination of persistent toxic substances from the Great lakes and zero discharge to control introduction of toxic substances into the Lakes.

A baffling problem, probably attributable to acid rain, is the disappearance of prolific frogs, toads and salamanders from wetlands that have been flooded or reclaimed. Some of the changing habitats causing the problem are changes in forest climates and streams caused by clearcutting, or a simple disorder in habitats such as vehicle traffic or cattle round-ups. Lawrence Cory, a Biology professor at St. Mary's College in Northern California, spent a decade in the 1960s studying mountain yellow-legged frogs whose home lies in an alpine lake in the Sierra Nevada Range. On returning to Koenig Lake 20 years later, these frogs had disappeared. Cory said, "It was the first time I had an emotional reaction to a change in the environment. I was shocked. It's like being in San Francisco at rush hours and not seeing any cars. Something's got to be wrong."

Acid rain's causes include emissions of sulfur dioxide and nitrogen oxides (NO_x). When released into the atmosphere, these substances can be carried by the prevailing winds and return to earth as acidic rain, snow, fog or dust. Sulfur dioxide emissions in North America come from coal-fired power generating stations and non-ferrous ore smelters, where the nitrogen oxide emissions are generated by vehicles and fuel combustion.

One of the most serious environmental problems in North America is caused by the acid rain producing serious economic, social and environmental problems. The increasing acidity of lakes and streams depletes aquatic life, such as the amphibians mentioned previously. Increases in acidity affect soil acidity, water and shallow groundwater. Other effects of acid rain include forest decline, erosion of buildings and monuments, and respiratory problems in people. In addition, acid rain endangers fisheries , tourism, agriculture, and forestry.

Canada and the U.S. share a transboundary acid rain pollution affecting both countries' lakes in a devastating manner. Within the U.S., six areas have lakes that are of great concern: New England, southwest Adirondacks, the Atlantic coastal plain, the upper peninsula of Michigan, the Appalachians and northern Florida. Canada has 300,000 vulnerable lakes with 150,000 being damaged and more than 14,000 are acidified. Additionally, 19 salmon rivers no longer support the species.

Sweden and Norway have an extreme problem with rain carried on prevailing winds from the European continent. By 1982, 18,000 lakes in Sweden became acidified and Norway's fish population declined in 2,600 lakes. In the Adirondacks, some lakes declined from a pH of 6.3 in 1933 to a pH of 3.9 in 1986. Since the pH is calculated loqarithmically, the increase in acidity is 100 times.

The geography and geology of lakes explains whether they can neutralize acid rain. The most vulnerable watersheds are those with rocks of granite, quartzite, quartz, and sandstone—these lack carbonates, bicarbonates or hydroxides that could neutralize the acidity of lakes. The midwest U.S. conversely, has bedrocks of limestone and dolomite capable of neutralizing acidity from acid rain. From the Midwest, prevailing winds carry acid pollutants to areas with low neutralizing capacities, such as the Northeast, Appalachians, Great Smoky Mountains, southern states, Minnesota, northern Wisconsin and Michigan, parts of the Rocky Mountains, Sierra Nevadas and Pacific Northwest. Vast areas of central and eastern Canada are affected, too.

As the pH of lakes falls, the fish find it more difficult to reproduce, especially in the spring when the snow melts and adds to the acidity. At this time, the young fish hatch from the eggs and cannot survive the new acidity. Healthy lakes have a pH of 6.5, but acid rain lakes with severe pollution have a pH of 4.7 or less.

How does the pH affect the aquatic life? Below pH 6 some species are killed, and below pH 4.7 nearly all species are injured or killed. Acid alters the body chemistry, impairs organ circulation, attacks the gills and interferes with the heart action. It affects the biological chain of life by destroying food species at the bottom of the chain, and thereby affecting all life forms from these to the top.

The Canadians experimented with an acid lake study on a nameless lake given the number 223. Scientists from the Freshwater Institute at Winnipeg led by Dr. David Schindler studied the lake in 1974, when its normal pH was 6.8. In 1976, the scientists started adding sulfuric acid each May, and by 1978 found the key organisms in the lake's food web started disappearing at pH 5.93—for example, crustaceans and opossum shrimp could not be found, anymore. Fat-head minnows, essential for trout feeding failed to reproduce.

By 1980, when the pH reached 5.59, the fat-head minnows were scarce, but the pearl dace, another variety of minnow, increased when their competition vanished. There were no young trout this year; in 1982 the trout spawned, but no young trout hatched from the eggs. By 1983, the pH neared 5.0 and no fish of any kind reproduced successfully. The trout were scavenged by other trout, and the remaining trout appeared skinny and misshapen.

Besides affecting aquatic species and the food chains, acid rain has scavenged Europe to a great degree. In the 1960s, acid rain was first recognized as a threat to forests. The first evidence surfaced in the Sudetes, a range of mountains important for timber production between Poland and Czechoslavakia. The fir trees showed thin branches and others were dying. The Germans were shocked to notice "Waldsterben", forest death, affecting their beloved forests, especially in the Black Forest and Bavarian Forest in the early 1970s. Early in the 1980s, the trees in 8% of the forests in West Germany were in poor health, and in 1985 more than half of the nation's trees were affected by forest death with an economic loss of $250 million annually. The German problem was closely related to nitrogen oxide deposition.

North America experienced the same problems with Quebec showing 40% of their sugar maples in decline. Quebec's maple syrup industry accounted for 89% of Canada's total maple syrup production, and was worth $33 million annually. Today, many of the 10,000 maple syrup producers have difficulty maintaining operations.

Eastern and southeastern U.S. report similar problems of forest decline. Many spruce and fir have died, or have dead crowns in the mountains of New Hampshire, New York's Adirondacks, Appalachians, North Carolina. In North Carolina's Mount Mitchell, the highest peak in the East at 6,684 feet, many spruce and fir trees above 5,000 feet have died. On Camel's Hump, a high peak in the Green Mountains of Vermont, more than half of the spruce trees died since 1965.

How does acid rain contribute to forest death? Although sometimes nitrogen and sulfur in acid deposition can actually fertilize some trees and plants in nutrient-poor soil, the effect eventually ends with a sharp drop in growth as nutrients are lost from soils leached by acidic compounds. Some lose their foliage, while others become damaged by insects, fungi and wind. The underground fungi become missing and scarce, and are vital to the well-being of plants. They aid in uptake of water and nutrients in the soil.

EFFECTS OF ACID RAIN ON HEALTH

Acid rain's damage to health hasn't been studied very extensively, but some medical doctors claim our knowledge about acid rain is as deficient today as the harm done by cigarette smoking was unknown in the 1950s. Dr. Richard M. Narkewicz, a pediatrician in Vermont and President Elect of the American Academy of Pediatrics in 1987 said, "the ingredients of acid rain (specifically ozone, sulfates and NO_x) cause damage in children and aggravate pre-existing respiratory conditions."

Pulmonary disease is responsible for 1/5 of all hospitalization of children under 15. Because their immune systems lack development, children have more acute respiratory infections. Dr. Narkewicz continues on the effect on children, "Children's airways are much narrower than an adults...A minor irritation caused by acid air pollution which would produce only a slight response in an adult— results in a dangerous level of swelling in the lining of the narrow airway of children... "

Doctors warn other segments of the population. Dr. Thomas Godar, President of the American Lung Association, says other people at risk from oxides of nitrogen and sulfur and ozone include people over sixty five; those with asthma, chronic bronchitis and emphysema, pregnant women and people with a history of heart disease. Dr. Philip Landrigan, Director of the Division of Environmental and Occupational Medicine at the Mt. Sinai School of Medicine in New York City says, "Acid rain is probably third after active smoking and passive smoking as a cause of lung disease."

Dr. Bonnie Stern of the Canadian government's Health and Welfare Department did a 1983 summer camp study by monitoring the respiratory health of 1400 children. She measured results by having the children blow into a machine measuring their lung capacity. Half of the group in a heavy acid deposition area in Ottawa "showed a small but statistically significant decrement in lung function compared to another group of children in an unpolluted area of Manitoba." The results of acid rain could surface later in life with chronic obstructive lung diseases like emphysema or bronchitis.

Besides the above health effects, acid rain destroys buildings, monuments, paintings, statuary, and painted surfaces of homes and cars. Acid rain either eats into surfaces immediately, or the dry acid deposition can settle into crevices and ledges and attacks later when the rain falls. It creates cracks and pits that make the stone vulnerable to water seepage or decay—causing microorganisms. Acid fallout causes "black scab" that damages the underlying paint and stucco.

An area not often mentioned in acid rain destruction is the Yucatan peninsula and southern Mexico. Many of the ancient Maya temples, murals and megaliths are being destroyed by the rain from the sky. Other historical monuments showing the effects of acid rain are the Taj Mahal of India, the Lincoln Memorial of Washington, D.C., the Arches of Constantine, Titus Septimus Severus, the Tower of London, Acropolis and Cologne Cathedral of Germany. The air pollution around Krakow, Poland will eat a tenth of an inch of stone from old buildings and sculptures in the next decade.

THE HISTORY AND POLITICS OF ACID RAIN

As early as 1921, some lakes in southern Norway had no fish population. By the 1950s, acid rain began its attack on the rivers and lakes of Scandinavia, and killed their trout and salmon. Later in the 1960s, the harmful effects of this air pollution spread to eastern Canadian waters and northeastern U.S.

During the 1960s, Gene Likens and F. Herbert Bormann of Yale University studied the acidity of the rain falling on an area of New Hampshire called Hubbard Brook. Likens says, "It was obvious that the rain was acidic, but we just assumed it was a local effect." The drama deepened when Likens went to Sweden on a fellowship, and spoke to Svante Oden and other researchers in the late 1960s. He suddenly realized that the acid rain problems of Sweden and New Hampshire were the same. When Likens moved from Dartmouth to Cornell University in New York State, he discovered more acid rain data originating in New York.

By 1972, Likens, Bormann and Johnson published a scientific paper warning the northeastern United States of the acid rain threat. His research became quite controversial, and stirred the wrath of the power industry who hired scientists to publish a paper denying the cause and effect of power plants on the problem. Fortunately, Likens colleagues supported the claims made by him and stated the reality that sulfuric and nitric acids originate in emissions from power plants.

At an earlier date in the early 1950s, Eville Gorham was a postdoctorate student in Sweden, who later returned to England and set up a research project in the lake district of northwestern England. He discovered winds blowing from the south from industrial Lancastershire or from the east from North Umberland carried rain incorporating sulfuric acid. He said, "I tried to work out the effect of daily weather conditions—how much rain fell and which direction the wind was coming from and so on. And of course, it was fairly easy to make the connection between acid in the rain and industry because when I got sulfuric acid the filters through which I was collecting my rain samples were sooty as all get out." Later, Gorham moved to the University of Minnesota, and discovered a critical area of acid rain in that state.

By the late 1960s, the pollution of the Great Lakes became a terrible reality. In 1978, Canada and the U.S. established a Bilateral Research Consultation Group (BRCG) on the long range transport of air pollutants. A year later, Canada and the U.S. announced their intentions to develop a cooperative agreement on transboundary air quality.

President Carter in his message to the Congress in 1979 called acid rain a serious environmental threat of global proportions. Under Ronald Reagan, the negotiations between the U.S. and Canada lost their momentum, and progress seemed to be stalled. Between 1981 and 1984 more than 3 dozen laws were proposed in the U.S. Congress to control acid rain, but negotiations were stalled by an anti-control alliance including representatives from states where acid rain originated. Other opponents were the coal workers, electric utilities, United Mine Workers, mining and auto industry and industry-supported research institutes.

In 1979 an agreement was reached with the United Nations Economic Commission Of Europe (ECE) in signing the Convention in Long-Range Transboundary Air Pollution. By the year 1981, the EPA Administrator concluded that acid rain damage from transboundary air pollution was occurring in both the United States and Canada, and initiated the international air pollution control provisions of The U.S. Clean Air Act. Dr. Bernhard Ulrich, a soil scientist who studied damaged beech and spruce forests in the Solling Plateau of West Germany for 2 decades, hypothesized that as the soil becomes more acidic, aluminum becomes soluble and toxic. The free aluminum attacks the tree's root systems.

Nine European countries met in an international meeting in Canada in 1984 reached an accord to reduce sulfur dioxide emissions by 30% by 1993. Later, the Helsinki Protocol of 1985 under the U.N. Convention was signed by 23 European countries and Canada with an agreement to cut sulfur dioxide emissions or transboundary flows by 30%. Three countries, the United States, Poland and Britain didn't sign.

Under President Reagan, the relations between the U.S. and Canada became strained as acid rain continued drifting to the north. By 1986, President Reagan and Canadian Prime Minister Mulroney endorsed recommendations to step up acid rain research. Canada launched an acid rain control program in 1985 to halve sulfur dioxide emissions in 10 years, Ontario worked hard to reduce four plants and smelters that emitted 80% of Ontario's sulfur dioxide.

The imminent acid rain danger took a step backward when the U.S. National Acid precipitation Program (NAPAP) reported that acid rain damage was neither wide-spread or worsening. The report was later dismissed as flawed, incomplete and misleading. The Sofia Protocol, under the aegis of the U.N. Economic Commission for Europe, Canada, the United States and 23 European countries, met in 1988 and agreed to freeze their NO_x emissions and reduce them to non-damaging levels.

A late Acid Rain Conference met in February, 1990, and reviewed the status of acid rain in the U.S. The acid rain problem is a combination of sulfur dioxide, nitrogen dioxide and other pollutants spewed from industrial plants and 130 million autos. The sulfur dioxide and nitrogen oxide emissions are decreased recently by the use of low sulfur coal, and scrubbers at power plants that produced 70% of the sulfur dioxide emissions in 1985.

Nearly 162 lakes in the northeast are acidic; this number could be cut in half in 50 years with a 30% reduction in acid rain. Varieties of trees affected by acid rain are the red spruce in northern Appalachians, a decline in sugar maples and reduction of pines in the Southeast. About 14% of the 250,000 buildings in the National Registry of historic Places in 1985 are constructed wholly or partially of stone and are susceptible to acid rain.

Some of the worst environmental air damage in the West comes from two coal-fired generating plants, the Laughlin Nevada's Mohave Generating Station and the giant Navajo Generating Station in Page, Arizona. Bob Yuhnke, of the Boulder, Colorado based Environmental Defense Fund works to reduce the haze in the Grand Canyon. He says, "We've recommended that the Mohave plant be the subject of a study similar to the Navajo Generating Station. The Mohave plant is the next largest source of sulfur dioxide in the West after Navajo."

The Mohave Generating Station was completed in the early 1970s before Bullhead City, Arizona and Laughlin, Nevada grew to a large size because of the booming growth of gambling casinos in Laughlin. At the present time, hundreds of thousands of people visit the area yearly, and 30,000 live near the plant.

A gray plume wafts from the Mohave Station and emits 37,380 tons of sulfur dioxide a year, second only to the Navajo plant. The combined sulfur dioxide fumes from both plants are a major source of white haze at the Grand Canyon, obliterating the Canyon from visitor's view. The Environmental Defense Fund in 1982 filed a lawsuit that triggered a study of the Navajo plant; it may be ordered by the EPA to install one billion in plant improvements.

Opposition to the EPA's plan comes from the plant manager of Southern California Edison plant in Laughlin. He will fight any attempt to force it to install costly emission controls, since the upgrading of power plants can cut hundreds of millions of dollars, and the cost would be passed on to electric consumers in California, Nevada and Arizona. Several people are looking at the plant's impact including Bill Burke, resource management specialist for the Lake Mead National Recreational Area and Mohave County Attorney William Ekstrom, Jr. He remarks, "I think if you can see it and smell it and it sits on your car on a calm day, then you know there's got to be something wrong with it. Eventually, within a few years, something will have to be done with the problem."

Recently, a happy ending to the Grand Canyon pollution story occurred. An agreement was reached to cut sulfur dioxide emissions from the Navajo Generating Station. Environmental groups had to take the government to court, and eventually reached an agreement to clean up the air over the Grand Canyon.

The final plan will reduce sulfur emissions by 90 percent by 1999. Under this agreement, some leeway will be given to plant owners, and will give them longer to install scrubbers, and will enjoy more flexibility in plant operations than under the original EPA proposal. After 30 days the law will be enforced under the EPA.

CLEAN COAL OPTIONS

Several different options are available to remove sulfur dioxide from coal that are environmentally and economically attractive. The least expensive coal refining appears in the form of coal washing where the coal is crushed and subjected to various purification procedures. Only a few power plants use this technology; it removes 25-40% of the sulfur from high sulfur coal. When done at the mine, the cleaned coal is less expensive to ship to power plants, and it produces less ash which can be an expensive proposition to dispose of.

The Electric Power Institute located at Palo Alto, California does research to explore other clean coal options. One of the most promising technologies offers hope to remove both sulfur dioxide and nitrogen oxides, and is called fluidized-bed combustion. The crushed coal mixed with crushed limestone burns above and mixes with the limestone. Limestone acts to absorb the sulfur from the coal, and produces calcium sulfate and calcium oxide which are removed from the furnace. The resulting ash is captured and subsequently discarded. Because of the low temperature of the furnace, most nitrogen oxides aren't formed in the process.

Coal gasification units are being employed by some electric power plants. The pulverized coal is mixed with water and oxygen to form a gas. The combustion of the gas drives the main turbine, while water is converted into superheated steam and runs another engine.

Another option to remove harmful sulfur dioxide fumes is called flue-gas scrubbing, and may be either "wet scrubbing", or a "dry scrubbing". Advantages lie in removing up to 90% of sulfur emissions from coal-burning plants, but the disadvantages are the enormous amounts of water used in the "wet scrubbing" process.

The by-products of a cake-like sludge must be buried in landfills, and the process fails to remove nitrogen oxide emissions. "Dry scrubbing" can use less water and energy, and is more economically viable then the "wet scrubbing".

A comparison of Canadian and U.S. technology in coal-fired power plants shows use of different technologies. Canada has two large coal-fired power plants, and they control their sulfur dioxide emissions with the use of low-sulfur coal and washed coal. The U.S. chooses scrubber technology for new coal-fired power plants over other means, both to meet emission regulations and protect miner's jobs in high sulfur coal fields.

SMELTING INDUSTRY CONTROLS

The production of metals is important economically to Canada, but causes an unhappy by-product in the production of acid rain pollutants. Most important ores such as zinc, copper and nickel occur as metal sulfides, and these sulfides produce sulfur dioxide during the smelting or refining of the metallic ores. Other by-products include volatized or vaporized metals (such as arsenic, mercury and lead), and particulate material which can promote metal contamination of the atmosphere. Some waste water is produced, but a process of neutralization removes heavy metals by precipitation.

The sulfur dioxide gas is released as "fugitive" emissions which escape from equipment during processing, or are discharged from the main smoke stack. The sulfur dioxide in the atmosphere can react with moisture to form sulfuric acid and sulfate compounds. The prevailing winds carry the compounds, and they eventually fall back to earth as acid rain far from their point of origin. Smelter's waste gas stream of over 6-7% sulfur dioxide can be converted to sulfuric acids. The conversion reduces sulfur dioxide emissions and produces a marketable product. These plants are called acid plants.

PROGRESS IN LIMING TECHNOLOGY

Sweden's liming program was designed to bring the acidity of thousands of lakes back to normal. From 1976-1986 the country spent $78 million to lime 4, 000 lakes, but the program was costly and acid deposition from other countries caused Sweden to consider liming a crisis intervention method. The Swedish National Liming Test program started in 1977, and the Norwegian Program began in 1981 and continued for 5-10 years.

The positive responses to their liming program showed successful reintroduction of fry, yearlings, one-year-olds, and mature adult fish into lakes; successful reproduction and recruitment of fish into lakes when prior to liming only older age fish existed, and the rapid growth rate for fish stocked into lakes with few or no fish.

Canada tried liming their lakes and rivers in Ontario and Nova Scotia, but encountered several problems. Liming is found to be a continuous problem of acid deposition which keeps occurring; when metal contamination exists, liming can not reduce metal concentration to non-toxic levels; it may not protect the habitat for critical developmental species in all fish, and in Europe the liming has led to domination by non-desirable fish.

In the United States, Living Lakes, Inc., a nonprofit Washington, D.C. organization has treated more than 28 lakes in nine states mostly in the northeast since 1986. Timothy Adams, Ph.D., environmental chemist and scientific advisor says, "Treated lakes require periodic reliming because the water is constantly being replaced. This is safe and viable method at least until a long-term solution is in place." It costs $10 to $25 per acre each year to test and treat a lake with finely powdered limestone.

The funds are contributed by coal and utility companies, and fish and wildlife groups supply labor and boats in spreading limestone. The Living lakes program was spread to Europe by a cooperative program between Cornell and Syracuse Universities and the

University of Freiberg in West Germany. The next chapter continues a problem closely correlated with acid rain destruction of forests, the physical removal of trees through deforestation. This removal of rainforests and ancient forests goes even farther in affecting the planetary air supply and demand.

9 EFFECTS OF FOREST ECOSYSTEMS ON AIR SYSTEMS

Ethnobotanist Paul A. Cox arrived in Western Samoa in July, 1988, and was distressed to find loggers attacking the rainforest surrounding the village of Falealupo. After a conference with the village chief, they told him there was no other alternative than cutting down the forest, since they needed the money to pay for the village's schools. They had no other money available to meet the required payments.

Surrounding Falealupa are 30,000 acres of the last surviving paleotropical rainforest on the island of Savai'i. Cox, who is an Associate Professor of Ethnobotany at Brigham Young University, worried about the rainforest destruction occurring for decades, and knows if logging continues for the next 20 years, the whole rainforest will be destroyed.

Cox contacted Forever Living Products and Nature's Way, both herbal product manufacturers, and they agreed to help the Western Samoan villagers with money to save their school. The donors required the high chiefs of the village to co-sign the covenant agreement to preserve the rainforest for 50 years. Meanwhile, the local people have a limited use of the rainforest for collecting medicinal plants, using certain woods and growing vegetables in small plots. Cox can continue his research of pharmacological rainforest products.

Why is there so much concern about rainforest and temperate forests? Trees are a very important part of the global air system, and should be treated with the utmost respect for their importance in mediating climate and air pollution. The exchange of gases between plants and the atmosphere are essential features of photosynthesis and respiration. Trees absorb carbon dioxide as an essential part of their photosynthetic processes, and release oxygen to the air as a by-product of photosynthesis. Currently, carbon dioxide is overwhelming the planet and causes the elevation of our planet's temperatures through the greenhouse effect.

Since the beginning of the Industrial Revolution 130 years ago, the combination of fossil fuels and deforestation have increased carbon dioxide in the atmosphere by 25% with a corresponding temperature increase. Trees absorb carbon dioxide at the rate of 48 pounds a year, or 10 tons per acre each year. Other large "sinks" for carbon are the Earth's oceans which absorb carbon dioxide, and its forests, especially tropical forest.

The forests can influence our climate by changing the chemistry of the Earth's atmosphere, storing large amounts of carbon, and having an important part in the hydrological cycle. Forests destroy the greenhouse gases and their precursors. In the tropical forest there are many hydroxyl radicals which are highly reactive chemicals which combine with the greenhouse gases, methane and carbon monoxide and are later removed by rain. If deforestation occurs, there might be a gradual increase in methane from both the lack of trees and from agricultural activities, resulting in excess greenhouse gases.

Our planet has a "sink" for excess carbon that can be found in Earth's ocean plants, land-based vegetation and the soils; this amount of carbon represents three times the amount of carbon that is stored in the atmosphere. Fossil fuel combustion and deforestation with associated burning of the tropical forests means we're playing with fire by changing our Planet's atmosphere content. Many forests and savanna grasslands in tropical regions are burned to create pastures and croplands, and this burning yields excess quantities of carbon monoxide, methane and nitrogen oxide.

Lastly, our climate is affected through the deforestation by a change in the water cycle. A study from the Amazon says evapotranspiration from trees recycles 6.5 trillion cubic meters of water each year in the direction of prevailing winds, and this amounts to one-half of the Amazon region's total annual rainfall. If deforestation of the Amazon occurs, a resulting problem would be a drastic reduction of the rainfall downwind. Global effects could include hotter and drier effects in the tropics and cooler temperate region conditions.

T.C. Pokhriyal writing in the "Indian Forester" believes forests can help control our pollution problems by reducing, but not eliminating air pollution. Some plants have materials on their exterior that can react with, fix or destroy air pollutants. Trees can remove hydrogen fluoride, sulfur dioxide, nitrogen dioxide and ozone from the air because they are readily absorbed. We should select trees that are pollution tolerant, resistant and insensitive to pollution.

Continuing T.C. Pokhriyal's research, he says the discharge of pollutants from large cities and industrial complexes has created problems that won't be alleviated by limited tree planting on the sites. Trees protect clean air by absorbing smokes and gases, and act as a windbreak to settle soiled particles. Effluents could be used to irrigate green belts or plantations, although some may need pre-treatment. Before trees are planted around industrial complexes, there must be a sufficient technical expertise. If the green belt remains effective, it must provide adequate safeguards against environmental hazards.

The well known cooling effect of trees is important in reducing the heat of urban areas. Global Releaf advises us that three properly planted trees around your home can cut your air conditioning bill by 10-15%. Much heat is generated from factories and cars stored in large areas of black-top pavement and by buildings. We can turn urban "heat islands" into cool and comfortable ones, and make the concrete jungle livable for man and beast. Global Releaf believes there are 100 million energy-efficient tree planting sites available around homes in towns and cities. Planting 100 million trees could offset America's carbon dioxide emissions by 18 million tons a year saving consumers $4 billion a year.

PAST HISTORY OF DEFORESTATION

Lessons from past history are of inestimable value in predicting problems of the future. Europe stands as a stark lesson of past misuse of forests, and these past events can help us to assess and prevent devastating forest destruction. For a thousand years after the last ice age ended in 8,000 B.C., Greece was covered with forests, but by the Fourth Century, B.C. the stark landscape and severe erosion caused these comments from the philosopher, Plato—"bones of a wasted body, the richer and softer parts of the soil having fallen away, and the mere skeleton of the land being left."

By the 16th Century in Europe, the demand accelerated for shipbuilding timbers and oaks became scarce around the Mediterranean area. When Philip II of Spain made plans for his Armada, he resorted to lumber from the northern European countries, since the supply of oak was exhausted. By the late 17th century, England needed to import timber from the colonies for their ship building requirements.

At this time, Americans started to spread West with resulting cutting of trees for timber and clearing of land for agriculture. By the end of the 19th century, very little of the North American woods were left. Our National Forests were established at the close of the 19th century to protect public forest lands from destruction by timber companies. At the end of this period, the loggers had stripped the Midwest and East bare with some resulting secondary and tertiary forests.

Currently, less than 10% of the virgin forest of our Northwest remains. It's been clearcut at the rate of 60,000 acres a year with hillsides scarred with logging roads, bare stripped and replanted with monoculture tree crops. The ecosystems of the Northwest Forests lays claim to over 200 wildlife species with the highest density of breeding birds of any temperate forests. The ancient trees are essential to this system and support over 100 plant species and more than 1500 species of invertebrates in their canopy.

We must save our own ancient forests, as well as help our global neighbors preserve their tropical rainforests. The newest research shows that tropical forests help to regulate our global climate and they are disappearing more rapidly than believed possible. A report prepared by the World Resources Institute and the United Nations, "World Resources 1990-91" estimate 40 million to 50 million acres of tropical forest, or an area the size of Washington state is disappearing.

James Gustave Speth of the World Resources Institute says, "We were startled to uncover this rate of global deforestation. We were saying we were losing forests at an acre a second, but it is much closer to an acre and a half a second." Eight tropical countries are contributing to this rainforest loss; Brazil, India, Indonesia, Vietnam, Thailand, Philippines, Costa Rica and Cameroon.

Why is everyone so concerned about the disappearance of the tropical rainforest? Besides its effect on the global climate system to alleviate the greenhouse effect and global pollution, there are other important facets of the rainforest. They are the richest, oldest and most productive and most complex of the ecosystems on the Earth, and cover 6% of the Earth's surface being home to 1/2 of the species of our planet. They are valuable in providing many of our foods and pharmaceuticals used in everyday life. We harvest coffee, nuts, honey, spices, and bananas from the tropics, and find they are home to wild our grain crops furnishing a gene pool of resistance to disease, insects and drought. One out of four medicinals come from tropical plants with drugs currently used to treat cancer, leukemia and Hodgkin's Disease.

Slash and burn agriculture contributes to rainforest devastation. Millions of acres of Amazon rainforest are slashed and burned to create farmland. Because the soil is thin in nutrients and ill-suited for growing agricultural crops, in a few years the farmers desert their land leaving it for cattle grazing and delve deeper into the rainforest. The landscapes become completely barren of rainforest cover. India's cutting of their rainforests is related to massive power plants designed to meet their energy needs, but the end result is rapid rainforest destruction and displacement of one million native people.

Slash and burn agriculture is unnecessary and economically feasible plans exist like the harvesting of latex for rubber. Anthropologist Dr. Stephen Schwartzman of Environmental Defense Fund helped rubber tapper leader Chico Mendes win approval from the Inter-American Development Bank for an "extractive reserves" program. The reserves consist of areas of the rainforest declared off-limits to bulldozers and destructive development, and are areas where the natives can live in harmony with the environment. There can be profitable harvesting of rubber, Brazil nuts and other rainforest reserves.

Mendes was tragically murdered in an ambush set by a cattle rancher on the evening of December 22, 1988. Opening the back door of his house, he was cut down by a blast from a 20-gauge shotgun in the Brazilian village of Xapuri. Dr. Schwartzman believes, "Now Chico has been gunned down by killers, men believed to have been hired by land speculators whose own selfish interests were threatened by his efforts to preserve the rainforest."

Dr. Schwartzman's letter continues, "The brutal irony is that most rainforest soil is too poor to sustain farming, or even cattle grazing for more than a few years ......so speculators and ranchers move on, clearing and burning more rainforest in an unrelenting march of destruction. And the rainforests that are lost......the oldest, most complex ecosystem on Earth.....can never be replaced."

Mendes achieved an international reputation and helped create a dozen extractive reserves in the Amazon. He earned the United Nation's highest environmental honor, the Global 500 award. Much more work needs to be done to save the rainforest of Brazil and retard the advance of the world's Multilateral Development Banks (MDB's).

Most of the destruction of tropical rainforest occurs because of Third World Development schemes usually financed by the U.S., European and Japanese taxes and private banks in the Northern hemisphere. The most destructive activities in Latin America are road building, logging, agriculture, mining, hydroelectric dams and cattle ranching. In Southeast Asia, logging and agriculture are the main culprits. The exploitation of tropical logging in Southeast Asia is due to bureaucratic institutions, transport infrastructure and investment capital. Teak became a great prize in the past, and was harvested in Dutch Indonesia, British India, Burma and independent Thailand. In the last 20 years, there has been a broader range of tropical species fall to local, American, European and Japanese axes.

An environmental organization, the Rainforest Action Network (RAN), recently boycotted the nationwide Burger King chain. How is this related to the rainforest? Two-thirds of Central America countries' rainforest have been cleared to raise cattle to produce cheap, stringy meat for American hamburgers. During the RAN boycott held in 1987, sales of Burger King hamburgers dropped by 12%. After the direct action, Burger King cancelled $35 million worth of beef contracts in Central America, and announced they would stop importing rainforest beef.

Deforestation effects on the atmosphere continues to be a tremendous liability and causes extra buildup of carbon dioxide. It can contribute an estimated 10-30% to the increase of carbon dioxide, can add to desertification and a general decline of ecosystems in drought-stricken Africa. The estimates of carbon released through deforestation range from 0.5 to 2.5 billion tons. On the other hand, efforts to reforest can result in an average of 2.5 tons of carbon being fixed or kept out of the atmosphere for every acre of trees planted according to Worldwatch Institute. Scientists at Oak Ridge National Laboratory speculate that a billion tons of carbon being emitted each year are missing, and say that possibly the missing carbon is being taken up through rapid photosynthesis by vegetation and plankton.

WORLD AREAS WITH DEFORESTATION PROBLEMS

Randall Hayes, director of Rainforest Action Network, says, "Rainforests are a nonrenewable resource. Once they're gone you can't reforest; the plants, animals and insects are driven to extinction. Just a few years ago, 14% of our land mass was tropical forest. Now it's down to 7%." A third of the world's tropical forest is located in the Amazon basin.

A situation within our United States boundaries exists in Hawaii where the destruction of one of our last native rainforests is threatened by a geothermal development. Randall Hayes believes a loss of American rainforest would undercut efforts to convince Third World countries Brazil, Malaysia and Indonesia to protect their rainforests. "How can we ask debt-ridden tropical countries to do what we in the U.S. aren't willing to do so?" RAN is currently boycotting the Hawaiian tourism industry to fight the geothermal project.

Another endangered area from the logging industry lies in the Pacific Northwest and northern California. In 1949, the cut from the entire National Forest System coast to coast was 2.6 billion board feet, but last year 16 billion board feet of public and private timber were hauled out of Washington and Oregon. This cut represents a line of log trucks more than 20,000 miles long. In the U.S., only 5% of the nation's original native forestlands remain in Oregon and Washington, less than 10% of the ancient old-growth forest still stands.

After an aerial tour by Lighthawk of Washington's Olympic National Forest, Dan Woods of "The Christian Science Monitor" said, "Spread out before us in a 20 mile square mile stretch of peaks and valleys that look like a Marine haircut gone astray; entire mountains shaved down to a stubble or bare dirt, leaving only the indent of logging roads spiralling to the top like threads of a corkscrew."

Chris Maser, scientist and former BLM biologist comments, "The ancient forest is our maintenance manual for the world's evergreen forests. Only we don't begin to know how many parts there are. The remaining ancient forests are our blueprint for the great coniferous forests which once covered this part of the world. This is the only place where the entire genetic code, the entire system is intact. If we liquidate the blueprint, we've lost it for the entire earth."

In six national forests in Oregon and Washington, only 33% to 50% of old growth sample tracts were still forested. Only 106,000 acres remain of 217,000 claimed by the forest service. In Oregon's Siskiyou, 142,000 acres remain of a claimed 433,000 acres. The old forest may only last another 20 years.

Earth First! environmentalists are tackling the tough job of saving the old-growth redwood forests of northern California by a non-violent summer of 1990. College students from 450 campuses across the country rushed to Mendocino County to participate in the protests. Betty Ball, coordinator of the Mendocino Environmental Center in Ukiah, California said she had calls from New York, Hollywood and Oregon and people are suddenly starting to realize the magnitude of the excessive logging in the old-growth forests. The trees at danger from logging companies include 1,000-year old redwoods and Douglas firs.

Some of the methods used in the non-violent protest include tree-sitting, tree hugging, log road blockading, banner hanging, and picketing, although Earth First! doesn't believe in spiking trees to thwart logging. All participants are given non-violence training, and must agree to a code that prohibits damaging people or property. A tape containing such songs as "You Can't Clearcut Your Way to Heaven", and "Knocking On Extinction's Door" incite the students to greater action.

The first event of the Redwood Summer was an effort to shut down Louisiana Pacific's log export dock in Samoa, south of Eureka in Humboldt County. The company exports logs sawed in half to qualify them as partly finished products to avoid the ban on export of raw logs from public lands.

The Northern Spotted Owl in the northwest forests, and the Mexican Spotted Owl in the Southwestern forests are suffering because of habitat losses from logging and natural disturbances. The owls are used by foresters as an indicator species to judge the health of the forest. The spotted owl is currently a threatened species. A government report recommended a management plan to end logging in some regions of the federal forests in Washington, Oregon and northern California. Scientists from the Forest Service, Fish and Wildlife Service, BLM and National Park Service proposed setting aside 8.5 million acres of forest that would reduce timber cuts by 25% in national forests and 30-40% in BLM forests. These recommendations are given only to protect the spotted owl and not the ancient forest issue.

Another region in the U.S. threatened from heavy logging is Alaska's Tongass with the largest National Forest in the U.S. This forest stands in great danger from a 1980 law passed in Congress to allow heavy logging in order to win support for Alaska National Interest Lands Conservation Act. This Act required the Forest Service to offer timber for sale totaling 450 million board feet per year which exceeds the forest's ability to regenerate itself.

The conservation group, Lighthawk, has shown congressional leaders and conservation leaders the results of the federal government's logging subsidies. The U.S. Forest Service spent more than $53 million in 1986 to build logging roads and promote commercial timbering, but only received $86,000 in total sales. Steele Wotkyns of Lighthawk said, "Trees that are over nine feet in diameter and 100 feet tall are sold for $3 each and turned into pulp for Japan's chemical and paper companies."

Many forestry experts believe the U.S. should point the way, and show the world that we can make better use of our forests, and use a reforestation plan to offset damages in denuded areas. Three countries, China, Gujarat in India and Kenya have made strides with successful programs in tree planting to offset the clearcutting and burning programs in other countries. In 1958, Mao Tse-Tung of the People's Republic of China proposed that each rural family plant a tree at each corner of their house. The tree selected for reforesting was the Paulownia which grows 8 feet/year, and is ready to harvest in a decade. Between 1970 and 1980, the tree planting tripled and produced an adequate supply for the country's one billion people. The people were so enthusiastic that trees found their way along roads, rivers, canals and boundaries between fields.

Although India has many problems with deforestation, a success story does emerge from the past. Gujarat was a tree-poor state in India; twenty years ago it lacked enough wood for their everyday needs, needing to import wood at high cost. In 1973, the government of Gujarat encouraged all the residents to plant wood lots, and let the local village councils and state share in the proceeds. Five years after the start of the program, one-sixth of the state's villages had new woodland plots with an emphasis on a Eucalyptus hybrid which grows 15 feet or more a year. This species has the ability to regenerate when cut down, and the leaves, bark and branches can serve as fuel. Today the villages are surrounded by unfenced wood lots.

Kenya shows great success with the Green Belt movement started by Wangari Maatha, a biologist for the University of Nairobi and Professor of Veterinary Medicine, who has worked with the rural women of Kenya. The Green Belt Organization has stopped desertification by planting millions of trees, and has convinced the African countries to start their own program. Green Belt helps the public to see the hazards of deforestation by targeting subsistence farmers who produce 70% of the staple foods. Deforestation makes their life more difficult by a constant search for fuel wood leaving less time to produce food. Currently, Green Belt has 600 tree nurseries, 10 million growing trees and 50,000 women and 3,000 schools working for its success.

In spite of the previous success stories, often areas of the world are facing destruction of the rainforests with the subsequent effects on our environmental air quality. The last large intact tropical rainforest stands on a 16,000 acre reserve, Wao Kele O Puna, on the Big Island of Hawaii. The entire ecosystem is threatened here by a geothermal power development of prospective twelve power plants. The True Geothermal Company of Wyoming plans to use the steam from underground waters and develop a 500 megawatt power plant. Native activists, the Pele Defense Fund, were bypassed by due process which enabled the state to set aside the rainforests for their own needs. The Wao Kele o Puna is a religious ground regarded as sacred by the natives and the volcano Mauna Loa is regarded as a manifestation of the goddess Pele.

If the geothermal project goes through, it will signal the onset of more industrialism, commercialism and tourism with condos and resorts. Non-native species could be introduced and a toxic by-product of hydrogen sulfide could be introduced into the atmosphere.

Rainforest Action Network is stepping in with a lawsuit to claim the land for native Hawaiians. Randall Hayes, Director of RAN, says, "If we allow these forests to be destroyed, it will be the most unforgivable of sins. We will be destroying tools of survival for future generations. If we save these forests, it will be our greatest accomplishment."

Although Hawaii's deforestation is a problem for our American states, the world's eyes are focused even more intently on the devastation being wreaked on the tropical rainforests of the Amazon, and the deforestation projects occurring simultaneously. The clearing is driven by both hunger and debt—the peasants are hungry and the politicians are in debt. Southeast Asia has sold most of its forests to Japan, and Japan is invading the Amazon. In Latin America, the numerous squatters lay claim to land by cutting down trees and "improving the land."

Integrally involved in many Third World Projects are the MDBs or Multilateral Development Banks and the International Monetary Fund (IMF). Every year the MDBs (World Bank, Inter-American Development Banks, Asian Development and African Development Banks) lend $25 billion to Third World countries to finance development projects such as roads, dams and electrical power plants. These projects wreak damage on the environment, and result in massive resettlement for the tribal people involved.

What kind of destructive projects have been financed by the MDBs? There has been massive rainforest clearing and agricultural resettlement in both Brazil and Indonesia, and a series of dams built in India resulting in more rainforest clearance, risks of water-borne disease, and dislocation of more than a million tribal people. Other projects resulting in deforestation are the clearing of land for cattle ranches, although local people can't afford the meat, but it is used solely for export.

One MDB success story mentioned earlier involved Chico Mendes and his efforts to start an "Extractive Reserve" for Brazil. Working with the Environmental Defense Fund, Mendes and the tribal people convinced the World Bank and the Inter-American Development Bank to finance the reserves for rubber tapping in the rainforest.

The Polonoroeste Project in northwest Brazil was financed by the World Bank at a cost of half a billion dollars, and 1/2 of this sum was used to pave highway 364 crossing the state of Rondonia. Proposed to relieve population pressure in southern Brazil, the eventuality became the rapid destruction of the rainforest through the migration of landless people. Thousands of these peasants cleared the tropical forests, but within several years the poor tropical soils became unsuited for agriculture. At this point, the people were forced to either sell their land to cattle ranchers or speculators, or deforest more tropical forests by slash and burn agriculture. Since 1980, the percentage of Rondonia covered by virgin forest has decreased from 97% to 80%.

Bruce Rich, an Environmental Defense-Fund attorney, has been extensively involved with the Polonoreste Project and the World Bank. Today, areas in the project suffer from illegal logging, mining and agricultural squatters. The public health situation of the Indians is poor with "constant epidemics of tuberculosis, measles and malaria".

Bruce Rich says, "The World Bank is now preparing two new Polonoroeste Loans totaling $350 million to support the ecological zoning and agricultural consolidation of the state of Rondonia and the northern Mato Grosso state. The Bank is not consulting with or involving the local population leading to mass protests over their dam building."

Other problems include the Bank financed Carajas Iron Mine and and Railroad. Plans to build pig iron smelters along the railroad threaten to deforest 58,000 square miles, an area half the size of Colorado and bigger than Pennsylvania. This is to occur in the next 20 years.

Bruce Rich says, "The World Bank needs to resolve many issues in Brazil. The Bank should insist that Brazilian authorities legally regularize the Amerindian Reserves and protected natural areas originally agreed upon. Consistently, there has been a lack of consultation with groups representing target populations and beneficiaries of the project; Amerindian and rubber tappers populations and agricultural colonist. There should be a physical establishment of seventeen Amerindian reserves and four protected natural areas, whose establishment was a consideration of the 1981 Polonoroeste loan."

"The agricultural component of the proposed new Polonoroeste loan is gravely flawed and inadequate. It should be reformulated to include economic analysis of smallholder cultivation of perennial crops of coffee and cacao."

The American Citizens can make a difference in destructive rainforest policy. We can influence MDBs, including the World Bank that receive millions of dollars from the U.S. and other industrialized countries to prepare economic plans and loans for development projects. The people of the Third World should have a larger voice, and be able to concentrate on domestic needs instead of being over-exploited for resources for export. We can apply legal and moral pressure to places like the World Bank to change their policies.

Another disastrous World Bank project is the Sardar-Sarovar Dam project in India which would flood 60,000 acres of forests and agricultural lands, with many in the Narmada's basin forested areas. It will ultimately result in the displacement of 90,000 tribal people, and small farmers and submerge 248 villages. Although the bank involved needs environmental policy statements, usually the bank fails to act in an environmentally sound way. This results in grave environmental and social problems. Lori Udall, an attorney for the Environmental Defense Fund says, "Four years after the project was approved, there is still no resettlement plan, or even a final count of the number of villagers that will be displaced."

The Sardar Sarovar Project doesn't consider the basin's natural resources and their climatic value— climate regulations, water conservation, soil conservation, forests and wildlife habitat and diversity. Eventually, the project will include 30 large, 135 medium-sized and 3,000 small dams on the Narmada River. The Environmental Defense Fund (EDF), Indian environmental lists and four U.S. environmental groups say the project should be cancelled. Lori Udall of EDF remarks, "Sardar Sarovar is an international scandal."

The World Bank seems to have good intentions, although not following through. In May, 1987, President Barber Conable delivered a speech in Washington. He vowed to create a new, greatly expanded environmental department, and to establish assessment units to monitor projects in four world regions. Twenty months later, there is a growing gap between the environmental rhetoric and actual field conditions. Problem areas continue to be equitable treatment and resettlement of "development refugees."

For several years, the Indian and non-governmental organizations (NGOs) have questioned the original cost benefit analysis of Sardar Sarovar, maintaining it underestimated or totally excluded some of the serious costs of the project. These costs include measures to mitigate public health projects, the full cost of R&R, the environmental and economic cost of forest submergence and the value of prime agricultural land in the submergence area of Madhya Pradesh.

The World Bank has admitted that the original/cost benefit analysis may be obsolete, and in meetings with the NGOs, the Bank has orally committed to revise the cost benefit analysis. It is believed the World Bank and Indian state governments did not adequately review the project. Additionally, India has a poor record for large scale irrigation projects—of 246 projects started since 1951, 181 are still incomplete.

Unfortunately, the World Bank and the State Government of Madhya Pradesh are conducting business as usual, and are appraising Narmada Sagar, the next major Narmada River Dam. Narmada Sagar will displace over 100,000 people and submerge 40,000 hectares of forest land and 44,000 hectares of agricultural land.

The Environmental Defense Fund believes the World Bank and state government should research and develop energy and irrigation alternatives that are less costly and environmentally and socially more sound.

ENVIRONMENTAL SUCCESS STORIES

Government policies and economic forces can be changed by teaching indigenous people how to make progress. Cultural Survival, an environmental group in Cambridge, Massachusetts, is an organization of anthropologists and researchers from many areas. They help indigenous people living in tropical forests around the world develop at their pace and with their culture intact. They establish markets for sustainably harvested rainforest products to increase the income of forest residents without destroying their forests or cultures.

An American ice cream company asked Cultural Survival to identify tropical forest products they could use in their ice cream flavors. The company agreed to accept 50,000 pounds of Brazil nuts and cashews harvested off the forest floor without harming the forests. If these nuts aren't harvested, they will rot on the forest floors. The group will deliver the nuts to The U.S. at a fair market price, and the rainforest residents will benefit financially from the transactions.

Economically, forests used to harvest Brazil nuts can produce five times more income per acre than the same land used to raise beef cattle. The World Bank estimates that it costs $50,000 to clear forests and create a single job in the cattle industry in Brazil. This same land producing Brazil nuts could supply 500 Indians with employment at harvest time and generate 1/4 to 1/3 of their yearly income.

The timber companies in Mexico were about ready to cut forest belonging to the Huichol Indians. With assistance from Cultural Survival, they built a carpentry/woodworking school. They now cut fewer trees, build furniture to sell locally and make 300 times more money per tree than received from the timber companies.

Another success story comes from the Arizona Rainforest Alliance(ARA), a grassroots organization dedicated to the preservation of rainforests and their inhabitants. Locally, they educate people about the extent, causes and remedies of rainforest destruction; internationally, they aid non-governmental organizations and indigenous people through monetary assistance and networking.

ARA sponsored a conference in September, 1989 bringing together representatives of the Lacandon Maya of southern Mexico, and representatives of native cultures from the U.S., including Tohono O'odham and Lummi. The Mayans reported the destruction of the Lacandon Rainforest, and requested the aid of Americans to help fight deforestation of their home. ARA donated money to the tribe to help preserve their way of life.

Perhaps their most successful campaign concerned the Altamira Dam Project in Amazonia, Brazil; it would have flooded a vast area of the Kayapo Indian Reservation. This dam would have resulted in the destruction of rainforest and the displacement of the Kayapo from their home. The project would have been funded by a consortium of banks, including Citibank. ARA worked with non-governmental organizations around the world, as well as Kayapo leaders to pressure Citibank to abandon the project. Public pressure forced Citibank to pull out of the project, which subsequently fell through.

Another organization taking an unusual approach is Lighthawk, a non-profit membership group based in Santa Fe, New Mexico. Their mission statement says, "We use flight research, political action and other technology to shed light on and correct environmental mismanagement. We foster conservation objectives and empower others to do the same thing through mankind's capacity to supporting ecological processes." They are an aerial conservation group that monitors pollution and deforestation in the U.S., Central and South America by flying conservationists, scientists and journalists over threatened areas.

Steele Wotkyns, Director of Development of Lighthawk, formerly with the Nature Conservancy says, "Our outstanding success story was a two-year effort brought about by the creation of a 97,000 acre nature reserve known as the Bladen Nature Reserve. We flew many government officials of Belize over the reserve and organized and sponsored ground expeditions. These yielded many high quality photos and video footage. Over a two year period, we negotiated directly with the government of Belize (formerly British Honduras). We gave them an incentive to promote this area for natural history tourism, and conserve this incredibly rich area."

Steele mentions working with Jim Jontz (D-Indiana) to make him aware of the forestry problems in the Northwest. Lighthawk took Jontz on an aerial tour over Oregon with the intent to show the problems with logging and clearcutting on the slopes with the subsequent erosion. Jontz became an active champion of Western forestry, and now thinks of the forests as a protected land not to be exploited by timber interests. Lighthawk flies many other scientists, politicians, environmentalists on aerial trips to show the devastation of our native forests.

The only flying operation of its kind, Lighthawk has flown thousands of missions, and served more than 100 conservation organizations since 1980. They own and operate two Cessna Turbo-210s—six seat, 200-mile-per-hour aircraft that can observe aerial conditions and do photography.

This group of pilots and volunteers named themselves after a mythical bird according to founder Michael Stewartt "whose purpose is to shed light."

Another program is the Trees for Tucson/Global Releaf. The American Forestry Association's Global Releaf campaign advocated planting 100 million trees in the U.S. to offset carbon dioxide emissions and the greenhouse effect. The Trees for Tucson/Global Releaf was sponsored by Tucson Clean and Beautiful, a non-profit corporation with the approval of the City of

Tucson, Pima County, Governor Mofford, Arizona Commission on the Environment and the Tucson Chamber of Commerce.

The City of Tucson's vegetation has decreased steadily as the city has grown; we have developed an "urban heat island" effect with the summer temperatures increasing 1° F each decade since 1947. This effect can be offset by planting 500,000 trees in Tucson to increase the tree canopy cover from 20% to 30%. The trees help to shade buildings, lower temperatures, to save energy for cooling, and will help reduce the excess carbon dioxide in the air. More oxygen for the atmosphere is produced by the trees, and they help to filter dust and particulates.

Part of Tucson Clean and Beautiful's program involves recycling to offset deforestation and use of trees. The organization sponsored the first telephone directory recycling program in the nation, sponsored by U.S. West Direct and the Circle K Corporation. Their recycling program diverted 225 tons of paper from local landfills and raised public awareness of recycling. For their efforts, Tucson Clean and Beautiful received a national recycling award.

The trees for Tucson/Global Releaf has a program to get citizens involved by planting desert-adapted trees that utilize low water. Also, there is a gift certificate program to give to others for birthdays, births, employee incentive award programs and marketing. The organization encourages school or neighborhood planting by concerned citizens.

SEVEN WAYS TO HELP

The Rainforest Action Network (RAN) developed seven simple ways to help the environment.

1. Choose domestic wood instead of tropical wood products. The rainforest falls when you buy mahogany, ebony, rosewood, or teak furniture. Use local woods of oak, pine, birch, maple or cherry.

2. Help protect indigenous tribes. Worldwide outcry helped to free Malaysian activists jailed for protecting commercial logging of world's oldest rainforest on Borneo. Still, five square miles are destroyed daily in Penan. Write letters of protest to government officials of countries involved.

3. Urge the World Bank to stop funding devastating rainforest development projects with your taxes. Send a letter to the president of World Bank urging him to stop financing rainforest dams and instead fund small-scale projects. Address:

> Mr. Barber J. Conable, Jr.
> President, World Bank
> 1818 H. St., N.W.
> Washington, D.C. 20433

4. Write your representative urging a ban on export of pesticides to Third World countries. Chemicals banned in U.S. and other countries are exported to Third World countries. The U.S. is trying to eradicate coca plants in Peru using toxic herbicides near Amazon headwaters killing rare rainforest plants and animals, and poisoning rivers and topsoils.

    Write your representatives.

    > The Honorable_____
    > U.S. House of Representatives
    > Washington, D.C. 20510

    > The Honorable_____
    > U.S. Senate
    > Washington, D.C. 20515

5. Stamp out Amazon fires. Last year NASA satellites spotted 170,000 fires used to clear rainforests in Brazil province of Rondonia. This area has lost nearly 20% of its rainforest

and is a rich ecosystem. Rainforest burning increases carbon dioxide which causes the greenhouse effect. Send a letter to the U.N. to request emergency planning to put out Amazon fires.

Write to:

> Mostafa Kamal Tolba
> Executive Director,
> United Nations Environmental Program
> P. O . Box 30552
> Nairobi, Kenya

6. Avoid fast-food hamburgers and processed beef products. The U.S. imports 120 million pounds fresh and frozen beef from Central America. Two-thirds of its rainforests have been cleared to raise cattle exported to the U.S. The beef is not labeled with country of origin. Write Secretary of Agriculture to enact beef labeling legislation.

> Clayton Yeutter
> Secretary of Agriculture
> 14th St. and Independence Ave., S.W.
> Washington, D.C. 20250

10 AUDUBON TO SIERRA

The first Earth Day celebrated on April 22, 1970 started the environmental movement, while the next Earth Day on Sunday, April 22, 1990 brought a huge citizen's army into committing their resources for environmental change. David Brower, once the president of the Sierra Club remarks, "This Earth Day has the potential to capture the world's concern for the planet and truly change the course of history." The environmental groups in the U.S. have risen from 976 to 18,000 different ones, a staggering number competing for membership. A survey taken about the quality of the air found 83% of the people are worried about the air pollution.

This chapter hopes to initiate the reader into the various types of environmental organizations, but can't begin to cover all of them. Once people identify with their hopes of accomplishment through joining an organization, then they will make a commitment and decide on the group that best fits their purposes. The organizations featured here have a purpose in the overall picture of the environmental air movement. Categories of organizations include those giving out information, lobbying groups, grassroots help, general and specifically oriented, such as nuclear and deforestation. Some deliberately challenge the Establishment like Greenpeace and Earth First!

Because of the potential for reaching millions of people, some of the history of the first Earth Day will be explained. Gaylord Nelson, the founder of Earth Day, 1970, realized there was a great interest in the environment from media broadcasts of oil spills, especially in Santa Barbara. Since campus teach-ins were commonly used 20 years ago, why not do a teach-in for the environment? After announcing plans for Earth Day, over 20 million Americans gathered for thousands of events in 1970.

Although some people argue this movement was a failure resulting in even more problems today, this awareness extended to Congress where they passed the Clean Water Act, The Clean Air Act, the National Forest Management Act, the Toxic Substance Control Act and others. The Clean Air Act worked for the first decade, but didn't forewarn of future pollution problems to come. Although autos are far cleaner today than in 1970, there are more people in the cities giving a potential for more use of cars since they live far from their workplace.

Perhaps, some of the ideas of Denis Hayes, the coordinator of the old and new Earth Day might be appropriate. Hayes is an environmental leader and ran Environmental Action twenty years ago; he later worked at the Worldwatch Institute, the Smithsonian and the Solar Lobby. Today, he is in private practice as a lawyer until he organized Earth Day, 1990. The Earth Day, 1990 Green Pledge covers many facets of environmental action. It reads:

> "Because our planet today faces severe environmental crises such as global warming, rainforest devastation, rapidly increasing population and water and air pollution... Because the planet's future depends on the commitment of every nation, as well as every individual...I pledge to do my share in saving the planet by letting my concern for the environment shape how I act: I pledge to do my utmost to recycle, conserve energy, save water, use efficient transportation, and try to adapt a lifestyle as if every day were Earth Day.
>
> I pledge to do my utmost to buy and use those products least harmful to the environment. Moreover, I will to the maximum extent possible do business with corporations that promote global environmental responsibility. I pledge to vote and support those candidates who demonstrate an abiding concern for the environment.

I pledge to support the passage of local, state and federal laws and international treaties that affect the environment."

Denis Hayes says, " I think that an environmentalist who thinks of the energy consumption of an appliance or automobile when buying it, eats lower on the food chain, recycles and puts a flow restriction on the showerhead is a better environmentalist than one who doesn't."

Before continuing on with other environmental organizations, perhaps the history of one of the oldest, the Sierra Club, may be an inspiration to environmentalists. It was founded on May 28, 1892 with 182 charter members and John Muir was elected its first president. John Muir found the wilderness, and especially Yosemite to be his life's vocation. Muir became for Americans what Kevin Starr calls a "prophet of all that the Sierra promised; simplicity, strength, joy and affirmation." "Muir upgraded the entire California relationship to the mountains. As a public figure, he set a standard of what Californians should be: challenged to beauty and to new passion for life by the magnificent land in which they found themselves."

Perhaps, Muir is best known for his fight for the primeval valley of Yosemite, Hetch Hetchy. He called it a "wonderful exact counterpart of ....Yosemite Valley...one of nature's rarest and most precious mountain temples." While Muir was fighting to create Yosemite as a national park, San Francisco officials planned to dam the Tuolumne River where it flowed through Hetch Hetchy. In spite of a tremendous wilderness fight for five years, the fight was lost to the government. The bill to allow the flooding of Hetch Hetchy passed in Congress in 1913 and the valley was lost.

From the earliest days of the organization to today, the purpose has been to explore, enjoy and protect the wild places of the Earth; to practice and promote the responsible use of the Earth's ecosystem and resources; to educate and enlist humanity to protect and restore the quality of natural and human environment; to use all lawful means to carry out these objectives. The Sierra Club has launched a huge global warming campaign and is actively interceding in deforestation issues throughout the world.

GENERAL ORGANIZATIONS

Other prominent environmental organizations can be categorized as general and a few of these will be spotlighted. Those I have selected as general organizations include the Audubon, National and International Wildlife, and the Izaak Walton League.

Although the Audubon brings thoughts of bird-watching to mind, the organization has become active in many areas and puts out the "Audubon Activist" newsletter. They call themselves the most effective grassroots, issue experts, legislation wranglers, sanctuary managers and wildlife managers in the country. Their letter features Action Alerts from Washington, D.C. when issues are urgent and access to Audubon's hotline which gives updates on environmental battles as they happen. Their hotline access includes a direct line to Capitol hill offices, the front lines of conservation.

They have fought to coordinate an Acid Rain Monitoring Network, have a wetlands watch where local wetlands are watched, monitored and defended, have action alerts through the Activist Network and have environmental bulletins or on-line computerized information for chapters and field staff. They furnish Audubon Adventures—educational material for teachers and students in 4th, 5th and 6th grades.

The National Wildlife Federation is a nonprofit organization dedicated to conservation education, creating and encouraging an awareness among the people of the world of the need for wise use and proper management of resources in soil, air, water forests, minerals, plant life and wildlife. The organization publishes beautiful magazines, "National Wildlife" and "International Wildlife", On Earth Day 1990, they sponsored a global warming project for college students called "COOL IT", and a Citizen's Action Network. They sponsor educational programs, such as National Wildlife Week, backyard wildlife habitats, camps and summits. Their grassroots activism extends to the 50 states and they are prominent in lobbying and litigating on national and state levels.

An organization that is very active in Acid Rain Control legislation is the Izaak Walton League of America based in Virginia. It was founded in 1922 to protect our nation's soil, air, woods, waters and wildlife, and has been lobbying on acid rain control for eight years. They specifically combat air and water pollution, foster stewardship of soils and private land and promote ethical outdoor behavior.

The Izaak Walton League is active in the following programs: Save Our Streams where they adopt, monitor and defend local streams; Wetlands Watch with local wetlands education and monitoring and defense; Outdoor Ethics Program with a national newsletter, clearinghouse, speakers on improving outdoor behavior; Acid Rain Control; Riparian Enhancement Teams—hands-on stream improvement on Western public lands and soil conservation and sustainable agriculture.

GRASSROOTS HELP

From general organizations, people might be very interested in organizations that promote active grassroots involvement. I will spotlight three conservation groups, Friends of the Earth, Earth Island Institute, and Nature Conservancy. Two groups with more radical means of grassroots activism are Greenpeace, and Earth First!

A clipping from "Tucson Lifeline" tells about Earth First! It doesn't compromise—we set forth the pure, hardline position of those who believe Earth comes first. They are emotional, passionate and angry with a sense of humor. They believe lobbying, lawsuits and letter writing and research papers are important, but not enough. Earth First!ers use confrontation, guerrilla theater, direct action and civil disobedience to fight for wild places and life processes. They defend tropical rainforests, end livestock grazing on public lands, oppose mining and in Arizona support reintroduction of wild species or save near-extinct species.

The Nature Conservancy works to protect rare plants and animals by protecting their ecological surroundings. They have a large international program to save rainforests and other lands outside the U.S. and create and maintain the largest system of nature preserves in the world. They would like to save 200 key tropical areas covering more than 100 million acres as parks and nature reserves through Latin America and the Caribbean. Every day at least three unique species of life disappears, and by the year 2000 the world could lose one plant or animal species every hour of every day.

Some of their unique campaigns are the Consumnes River in California where they're replanting 100 acres of rare valley oaks and are expanding and restoring a 3,000 acre wetlands oasis. In Big Bend Coast in California, they bought 68,709 acres linking a 200 mile stretch of potential public lands with critical coastal habitat for manatees and Kemp's Ridley, imperiled green sea turtles and black bears, bobcats and woodstocks.

David R. Brower seems to be a link with the next two organizations, as well as the Sierra Club. He worked with the Sierra Club, Friends of the Earth and is currently the founder of Earth Island Institute. One of the most influential 20th century environmentalists, he was the Sierra Club's first Executive Director in 1952, founded Friends of the Earth, and now works with Earth Island Institute.

The Friends of the Earth merged with the Environmental Policy Institute and the Oceanic Society. They act as ears and voice for scores of grassroots organizations that bear the direct burden of environmental abuse. They are specifically interested in Agriculture and Biotechnology, Central America Toxic Chemicals Safety, Coal mining, Coasts and Oceans, Global Warming, Groundwater, Ozone Layer Depletion, Nuclear Weapons Development and East/West relations.

Friends of the Earth publishes a quarterly newsletter on protecting the ozone layer, "Atmosphere". Right now, they are campaigning for government regulations to ban the sale of CFC refrigerants in small cans to charge leaky air conditioners in cars, and to require the use of CFC recycling equipment at all service shops. Their campaign urges stronger governmental regulations and a ban on the 14-ounce CFC cans.

The newest David Brower organization, Earth Island Institute, has its hands on over twenty separate projects, many of them at grassroots level. They work to build ecological conscience and local citizen action. He says that the Earth badly needs CPR—Conserve, Preserve and Restore the life support system.

- Conserve—to budget resources wisely and equitably

- Preserve—to save the wildness that holds the answers to questions we will someday learn how to ask.

- Restore—to heal carelessly damaged resources—forests, waters, skies, wild things, cities, schools, health, human dignity—as best we can.

Earth Island first opened its office in 1985 in San Francisco, and provided for the Brower fund, "Earth Island Journal", Environmental Project on South America, International Marine Mammal Project, International Rivers Network and Rainforest Action Network. Their work on the Environmental Project in South America, EPOCA, alerts tens of thousands to the root causes of environmental destruction in Central America which includes environmentally destructive U.S. military aid, toxic waste dumping, clearcutting the rainforests, and wars against the people and environment.

An example of one of their projects was the peace trees planted in Central America. Fifteen young people from the U.S.S.R. and fifteen from the U.S. joined fifteen Costa Rican to plant 2000 "Peace Trees" in Bri Bri, a village near the northern Atlantic Coast. The students took courses on deforestation in the tropics and possible solutions to the problems. The friendships they formed during two weeks of living and working side-by-side will serve as examples of international cooperation.

The last militant grassroots organization to be covered here is Greenpeace. Greenpeace believes that one person can make a difference, and you don't have to be rich, famous or have powerful friends. Greenpeace was founded on "bearing witness", the courage to act, creative solutions and cooperative support. When you actively see an injustice being done, you must accept responsibility for that injustice. This requires courage, such as guiding a small inflatable boat near the hull of a giant freighter as barrels of radioactive waste are dumped, or plugging the drainpipe of an industrial polluter spewing toxic chemicals into a waterway.

One of their creative solutions resulted when they sailed their boats into nuclear weapons testing zones in Alaska and the South Pacific and put an end to nuclear testing in the atmosphere. Greenpeace boasts of 4 million supporters with offices in 23 nations around the world. They have a fleet of 7 ships, 50 inflatable boats and 2 campaign buses. They are an international environmental organization qualified to handle toxic pollution, global warming and the killing of whales and dolphins.

LOBBYING ORGANIZATIONS

Many of the organizations already covered are lobbying groups including the Sierra Club, Audubon Activist, Izaak Walton League, National and International Wildlife Organizations.

However, there are several featured in this section whose primary focus is lobbying.

The parent organization of Critical Mass (a nuclear organization) is Public Citizen founded by Ralph Nader 1971. It features citizen research, lobbying and litigation and fights for consumer rights in the marketplace, for safe products, for a healthy environment and workplace, for clean and safe energy sources, and for corporate and government accountability. Because it doesn't accept government or corporate grants, it is able to challenge the antidemocratic influence of special interest lobbies and political action committee. Special affiliate groups include Buyer's Up, Congress Watch to monitor legislation on Capitol hill, lobby for consumer interests and organize activists. Other affiliates are the Health Research Group, The Litigation Group to bring lawsuits on behalf of citizens against the government and the Tax Reform Group.

The Environmental Defense Fund is active in many air quality issues, and is dedicated to developing and promoting creative solutions to environmental problem through a partnership of science and law. Attorney Robert E. Yuhnke of EDF's Rocky Mountain Office is a leader in a clean air campaign for his home state of Colorado, as well as for acid rain problems caused by two power plants in the West, the Mojave and Glen Canyon. They have just started a strong campaign to stop the the destruction of rainforests in Brazil and elsewhere.

The EDF is working to stop the Multilateral Development Banks (MDB) from lending billions of dollars to projects that destroy tropical rainforests and other natural resources. The World Bank responded to over 20,000 petitions from EDF members to create an environmental department to consider the environmental impact of proposed development. EDF is attacking global environment by examining worldwide energy and agricultural activities that produce greenhouse gases. They would like to find an international, legal and political framework to limit the destructive impact of these activities.

An example of a local "watchdog" group is the Arizona Center for Law in the Public Interest which fights for protection of air, water and natural areas of Arizona. Their chief concern is to provide full-time legal advocacy for the state. Examples of bills they are working on are the regulation of industrial emissions of toxic chemicals into the air, requirements that environmental impact analyses be completed in all major publicly funded projects in the state and an increase in the level of fines assess against polluters.

They were triumphant in the Ninth Circuit Court of Appeals to clean up Arizona's unhealthy air. The court ordered the EPA to crack down on air pollution in Tucson and Phoenix. Oak Creek near Sedona was threatened with contamination from improper waste disposal. The Center forced the adoption of tougher standards for wastewater systems in Sedona.

GROUPS WITH SPECIFIC PURPOSES

I discovered a number of specific groups organized around nuclear energy and radiation problems, and around deforestation and tree issues of replanting. The Union of Concerned Scientists (UCS) has campaigns for global warming and cool energy, or alternative fuel sources as well as emphasizing nuclear and radiation problems. The UCS is composed of scientists and other citizens concerned about the impact of advanced technology on society and focuses on national security policy, nuclear arms control, nuclear power safety and national energy safety. It was originally formed as an informal group of MIT scientists, faculty and students.

UCS features UCS Action Networks where scientists participate in grassroots educational and lobbying programs, legislative Alert Networks, and a Professional Coalition for Nuclear Arms Control. They frequently provide expert testimony and analysis to Congressional Committees and have a UCS Speaker Bureau. They were the first to make extensive public examination of nuclear power—the 1971 Atomic Energy Hearings which exposed weaknesses in nuclear plant system design. They helped to make major safety improvements at power plants, assessed renewable-energy alternatives and helped on emergency planning and design issues at Three Mile Island and Indian Point reactors.

The Critical Mass Energy project is an offshoot of Ralph Nader's Public Citizen. Their 1989 program hoped to promote sound energy options for global warming and oppose the nuclear industry's attempt to resurrect nuclear power. They would like to alert the public to energy conservation, renewable energy systems and natural gas technologies to displace coal and oil at lower environmental and economic cost.

They continue to work with local safe energy groups to close the nation's nuclear reactors, and continue to publicize and correct safety problems at the nation's nuclear plants. They will expand to include nuclear waste problems and oppose creation of "low level" waste dumps, and oppose building a high-level waste dump at Yucca Mountain in Nevada. The number of activists working on commercial nuclear power issues has dropped 50% during the 18 months preceding 1989.

An unusual approach to the problem of nuclear weapons testing is given by the Nevada Test Experience. It's a faith-based organization with a Franciscan origin and scriptural values working to end nuclear weapons testing through prayer, dialogue, and non-violent direct action. NDE is a voice in the desert calling people of faith to nonviolence in the face of violence, truth in the face of illusions, hope in the face of despair and love in the face of fear.

In the 1970's, concerned Franciscans started holding prayer vigils at the Nevada Test Site. Others of different religious faith joined them for vigils and actions during Lent of 1982. After being organized in 1984, it offers the August Desert Witness to commemorate Hiroshima and Nagasaki. This organization uses nonviolent direct actions similar to Gandhi's through peaceful demonstrations, vigils, prayer and civil resistance. The action is specifically directed to the Nevada Test Site located 65 miles northwest of Las Vegas in Nye County operated by the U.S. Department of Energy.

Another air problem associated with global warming concerns the rapid deforestation of areas of the world including the rainforest region. Four organizations already mentioned have far-reaching deforestation campaigns, and these include World Wildlife Fund, the Environmental Defense Fund who worked with Chico Mendes, the slain Brazilian leader, the Sierra Club whose campaign will be emphasized in the deforestation and the Audubon Society, who are working to save the ancient forests of the Northwest.

Perhaps the oldest organization is the Global Releaf, a program of The American Forestry Association which was founded in 1875. This association helped Congress to pass laws creating the National Forest System and Forest Service, and established the National Park System and state forestry agencies. Global Releaf is a national campaign to mobilize Americans to plant millions more trees, and support forest management efforts in their communities. The campaign promotes policies to conserve energy and improve forests around the world.

They recruit Global Releaf Cooperators who agree to make a contribution and then supply the Cooperator with free material to spread the news. This material includes Global Releaf brochures, Citizen Action Guides, Tree Planting Guides, Urban Forest Facts and a free Arbor Day/Earth Day kit. They emphasize that Global Releaf is a grassroots movement dependent on people like garden club members, teachers, service club leaders, nursery or environmental advocates to help their campaign.

I recently received information from The National Arbor Day Foundation and joined it when I found that my small contribution would give me ten trees to plant that are adapted to my area. In addition, there is a subscription to their magazine, bargains in America's finest apples from their J. Sterling Morton Orchard, members-only weekends and saving 33% on the purchase of any kind of tree including fruit and nut trees.

TreePeople is a Los Angeles based environmental group which works with its partner, Global Releaf. The staff and volunteers of TreePeople have been planting drought resistant, smog tolerant trees in the mountains around Los Angeles. It had a successful campaign to plant a million trees before the 1984 Olympic Games. The organization works through planting coordinators and Citizen Forester groups, and conducts education through speakers, environmental fairs, events and festivals.

Within the U.S. in Eugene, Oregon, the Native Forest Council works to preserve the nation's last remnants of native virgin forests. It opposes the clearcutting of the original, untouched forests that existed when the first settlers came to America. Many of these forests are in public land and we are entrusted to preserve these forests. Unfortunately, our public resources are being exploited through pro-timber industry appointees, and the U.S. Forest Service and the Bureau of Land Management.

Besides our own forests, the rainforests of the world are in grave danger. Several groups are actively working to stop this destruction including the Rainforest Action Network (RAN), the Rainforest Alliance and the Rain Forest Foundation. All of these groups are appalled at the destruction of the rainforests with species extinction, rapid loss of biodiversity, and the broader effects on global issues.

For example, Malaysia is the world's leading exporter of tropical timber, and their logging industry is dominated by the Japanese timber carte, Nanyogai Freight Agreement. Because of this cartel, the rainforests in the state of Sarawak are being logged at seven and one-half acres every minute. Within 10 years, the Malaysian rainforests will be bereft of valuable timber species. Other nearby tropical nations running the same risk are the Philippines and Indonesia.

INFORMATION CENTERS

Several clearinghouses for information exist, but the listing is not comprehensive and only involves a sample of information centers. The Center for Environmental Information was established in Rochester, New York in 1974 as an answer to finding timely, accurate and comprehensive information on environmental issues. The CEI publishes a number of periodicals such as an environmental newsletter, directory of environmental organizations and technical digests on acid rain and global climate change. It, also, sponsors conferences and seminars on vital issues and provides information specialists who have access to worldwide databases through their computer systems. They work specifically on acid rain and global climate change.

A corresponding information center for nuclear issues is The Nuclear Information and Resource Center in Washington, D.C. It's a national clearinghouse and networking center for people concerned about nuclear power issues. It provides a quarterly journal with an in-depth analysis of major nuclear issues and "The Nuclear Monitor" for investment community. Citizens Guides to the aging of nuclear power plants are provided, and packets on radioactive waste problems. Radiation and health packets can be ordered, and also packets on advanced reactors, global warming and alternative energy.

The last information center with global implications is the Worldwide Institute based in Washington D.C. It was founded in 1974 to inform policymakers and the general public about the interdependence of the world economy and its environmental support system. The research staff analyzes issues from a global perspective and is an interdisciplinary framework.

They publish Worldwatch papers, an annual "State of the World" report and "World Watch" magazine. These publications are used by government officials, journalists, economists, business leaders, development specialists and environmentalists and students in every country.

11 GOVERNMENTS AND POLITICIANS

The United Nations Environment Program, UNEP, started in 1972, was the first UN program based in a Third World Country, Kenya, Africa. The UNEP serves as a catalyst and coordinator with governments, international organizations, scientists and environmental groups. Many of its most important issues concern environmental air problems, such as atmospheric pollution and climate change, controlling hazardous wastes and the potential loss of one million species.

UNEP believes if present trends continue for ten years until the year 2000, erosion will occur on one-third of the world's productive land; the trend will lead to a world-wide climate change causing annihilation of a million species. In the decade of the 1980s, many disasters happened. The world experienced the Bhopal leak from a pesticides factory in India, the Chernobyl nuclear reactor explosion with nuclear fallout increased risks of future human cancers, and a liquid gas explosion in Mexico City killed 1,000 people and left many thousands without homes.

UNEP plans to slow the arms race to release financial, material and intellectual resources, and would like to reduce the $1.3 trillion debt burden on developing countries. This UN body claims it is impossible to separate economic, issues from environmental issues, and poverty is a major cause and effect of global environmental problems.

UNEP monitors all atmospheric problems including acid rain, ozone layer loss, global warming, and is forming monitoring organizations for tropical forest loss. UNEP uses a 50-nation network to watch changes in chemical pollution in the atmosphere. The 1979 Convention on Transboundary Air Pollution affected the problem of acid rain by a signed agreement with Canada and 17

European countries. In 1989, 81 governments and the European Economic Community agreed to completely stop the production of CFCs by 2000, and to phase out other ozone-depleting substances.

The greenhouse effect is a necessity for life on earth, but a negative effect is the pouring of too much carbon dioxide into the atmosphere. In November 1988, UNEP and the World Meteorological Organization formed the Intergovernmental Panel on Climate Change. They will try to unify the scientific and policy-making communities for action on climate change, and hope a treaty may be ready to sign in 1992.

UNEP tries to save the world's tropical forests which are disappearing at a rate of nearly an acre a second or 3,000 acres an hour. These tropical forests cover 6% of the Earth's land surface, and are a source of 1/2 of our medicines, help to regulate the climate and prevent floods and landslides. Agriculture, cattle ranching, logging, highways, mining and hydroelectric dams cause the loss of these forests. UNEP introduced the environment into the International Tropic Timber Agreement, and also implemented The Tropical Forestry Action Plan.

Earthwatch helps to assess the global environment, and disseminates information in 142 countries. One major part of Earthwatch is the Global Environmental Monitoring System (GEMS) coordinating all national, regional and global data, and monitoring renewable resources, climate systems, air, water and food contamination, pollutants that cross national boundaries, rare and endangered species and parks and protected areas. Two other parts of Earthwatch are the International Register of Potentially Toxic Substances (IRPTC), and the International Program on Chemical Safety (a part of the International Labour Organization and the World Health Organization).

The IRPTC provides scientific, technical and legal data to help developing countries explore problems of 80,000 chemicals on the market and 2,000 new introductions each year. It collaborates with the other chemical organizations above to issue guidelines for exposure and testing, and reduce the damages of pesticide use in developing nations.

The UN sponsors some new conferences on various air issues, and held one in June, 1990 at London on the ozone layer. Susannah Begg, an Australian teenager representing the first youth delegation ever to attend an ozone conference said, "We have been watching you. It has been at times fascinating, at times confusing, at times horrifying ...We have had to keep reminding each other that what is actually being debated here is the future of the ozone layer. This debate has been largely guided by short-sighted commercial gains and national self interests...Your pursuit of diplomatic compromises is compromising our future...Remember that we will inherit the consequences of your decisions. We will not sign the Montreal Protocol—you will. You will not bear the brunt of ozone depletion—we will. We demand that you think in the long term...Our fate is in your square brackets."

The London ozone conference was signed by the European community, 53 of 57 countries that ratified the 1987 Montreal ozone treaty, and was approved by 40 other countries who participated but couldn't vote. An international ozone defense fund with a three year budget of $160 million was established to help convert factories that produce CFCs in developing countries to more ozone-friendly chemicals.

The UN is planning a 1992 Conference on Environment and Development among the world's nations. Some goals set for this new conference are:

- Specific legal measures to protect the atmosphere and biological diversity.

- Develop a statement of principle as a basis for conduct by nations and their people to ensure the future integrity of the Earth as a hospitable home for humans and other forms of life;

- Create an action agenda—"Agenda 21"—that will establish priorities, procedures, targets and responsibilities for the international community beyond 1992 into the 21st century;

- Implement "Agenda 21" by providing resources (e.g. money) and access to technologies, particularly for developing countries, and by strengthening institutions and processes.

INDIVIDUAL NATIONAL PROGRAMS

Some nations document great success reducing air pollution. Canada's progress was explored at length in the acid rain chapter. In March 1985, Prime Minister Brian Mulroney announced a comprehensive acid rain program. The threefold objective said that the Canadian house should be in order if Canada expected The U.S. take action, the Canadian emissions reductions would reduce some acid rain damage occurring in Canada, and Canada should be responsible to eliminate any damage it contributes to the northeastern U.S.

Their controls continue right on schedule, and record notable achievements. The seven easternmost provinces show a reduction in sulfur dioxide emissions of 40% (from 1980 levels). The entire nation achieved a 33% reduction on the flow of sulfur dioxide from Canada to the U.S. The Clean Air Act program in The U.S. should reduce U.S. sulfur dioxide by 10 million tons (from 1980 levels) by the year 2000 with one half of the reductions in place by 1995.

The combined benefits of the U.S. Clean Air Program could reduce transboundary flows of sulfur dioxide by 25% by 1995 and a full 50% by 2000. Canadian experts project that combined efforts of the U.S. and Canada would decrease eastern Canada's acid deposition by 1995 to 90% less than it was in 1980.

Mr. Charles Caccia, Minister of Environment Canada, said at an acid rain conference in 1984, "Acid rain poses a threat to the basic economic resources of my country: forests, lakes, rivers, fish, agriculture and wildlife.....Canada's forest industry is a major contributor to our economy, employing one in ten Canadians directly or indirectly. Shipments of forest products from Eastern Canada amount to about $15 billion a year. Of the lakes surveyed in the province of Ontario, 43% are vulnerable to acidification... The salmon fishery in our Atlantic provinces is suffering the impact of acid rain. In Nova Scotia, salmon no longer run in nine former salmon rivers, and there are initial signs of acidification in twice as many again....."

Sweden and Norway decided to clean up their act before asking for help from Western Europe and Great Britain. Although NO_x remains a problem, Norway cut sulfur dioxide emissions by 30% from 1976-1986, and Sweden's sulfur dioxide reduction was cut 60% in the same years by switching to low-sulfur fuels and nuclear fuels. The Swedish government is ahead of the U.S. in their education and research programs, and it publishes a full color magazine called "Acid". (Box 1302, 3-171, 25 SOLNA, Sweden).

Mr. Svante Lundkvist, Minister of Agriculture of Sweden at the Acid Rain Conference of 1984 said, "Recognizing that our forests —a very important perhaps crucial asset to the Swedish economy—are showing signs of stress caused by too much pollution, we are also aware of the continuing acidification of our soils, lakes, running water and ground water. 18,000 lakes in Sweden are today, to a varying degree affected by acidic substances."

In West Germany, more than half of the nation's trees were affected by Waldsterben (forest death), and the German timber industry estimates a $250 million loss from Waldsterben. A large percentage of acid deposition comes from nitrogen oxides and they plan to lower speed limits on their autobahns, and require cars built after 1986 to have catalytic converters and use unleaded gas.

At the 1984 Acid Rain Conference, Mr. Carl-Dieter Spranger, Parliamentary State Secretary of West Germany, Federal Ministry of the Interior said, "The ecological consequences of damage to forests and waters threaten our natural basis of life. In this context, I would like to point out the ecological importance of forests for the water balance, climate and protection of soils as habitats of plant and animal species. The damage to forests, waters and buildings is causing an annual loss of several million DM to our national economy. At present, affected forests alone are estimated to account for losses of more than one thousand million DM per year in my country; more then 1 million persons working in tourism, forestry and wood processing industries depend on forests to make their living. Irreversible damage is done to valuable historic monuments and works of art....."

Alan S. Miller, the Executive Director of the Center for Global Change at the University of Maryland spent seven months in Japan as a Fulbright Scholar studying Japanese perspectives on global environmental issues. He believes Japan has made substantial progress in cleaning up domestic pollution, and in some areas such as recycling and urban air quality, now betters the U.S. Because Japan, now the second largest economic power, brings added international responsibility to its success, it is important to follow its progress. Its success impacts the environment, e.g., more than 40% of Japan's wood imports are from tropical countries, and Japan uses more than 10% of the annual global total of chlorofluorocarbons which damage the ozone layer. Even though they are noted for energy efficiency, they still are the world's fourth largest source of carbon dioxide emissions.

On the financial side, Japan has ten of the world's largest banks, and 53 of the 100 largest companies. Because of their growing wealth, they contribute financial aid to developing countries, and are the second largest contributor after the U.S. They favor commercially oriented trading partners in Asia, and low-interest loans rather than grants, and tied rather than untied aid.

Japan supports the U.N. environmental activities; proposed the establishment of the World Commission on Environment and

Development, and gives environmental and technical assistance to Third World countries. The Environment Agency of the government of Japan proposed four new policy recommendations. They should ratify the Montreal protocol on Substances that Deplete the Ozone Layer; secondly they should strengthen environmental aspects of foreign aid, increase the commitment to scientific research on global environmental solutions and develop technologies. Lastly, non-governmental organizations and the public need to become more involved in the environment. In Japan's favor, all of these recommendations apply to worldwide nations involvement.

Since the majority of the readers of this book are located in the United States, we need some success stories from the U.S., as well as global environmental efforts. The South Coast Air Quality Management District of Los Angeles was featured extensively in the Toxic Air chapter, but some of their general policies are worthy of further examination. The AQMD seeks to achieve and maintain healthy air quality through a program of research, regulation, enforcement and communication.

Because Southern California has the worst air pollution in the nation, their rules provide firm guidelines to reduce all air pollution from factories, power plants, oil refineries and chemical plants. Los Angeles follows the federal Clean Air Act standards, while adopting specific strategies and control measures to achieve these standards. Later the EPA and California Air Resources Board reviews current strategies, and the District formulates rules to put the plan into practice.

These steps are used in making a new rule.

1. The idea comes from the District Staff, Governing Board, from federal or state agencies or from the public.

2. The District Rule Development Divisions does research to find out who would be affected by the rules, and how and

how much pollution can be cut. The staff writes a draft of
the rule.

3. In an Internal Workshop, the Rule Development staff
   reviews each proposal with representatives including
   experts in planning, engineering, enforcement and technical services.

4. Rule Development mails a notice of the draft rule to public
   associations and companies that would be affected. They
   ask for comments on the proposal and invite interested
   parties to a Public Workshop.

5. After considering input from public and businesses
   involved, the staff incorporates ideas into a Proposed Rule
   and Staff Report. After public notice, there is a public
   hearing.

6. The Governing Board must consider input of all parties
   including public, industry and staff before voting on new
   rule. Then enforcement happens if the Board votes for the
   rule.

Besides the results achieved by local governments, such as the AQMD of Los Angeles, there are a number of private organizations associated with universities; two notable ones are The Center for Global Change at the University of Maryland, and the Morris Udall Center for Studies in Public Policy at the University of Arizona. They maintain active research scientists, politicians, lawyers and communications experts to influence public policies. They often sponsor global warming seminars, and other pertinent topics in an effort to influence the public of the potential threat of these problems.

The Udall Center, christened after the Udall family, represents a commitment to public service and responsive public policy. Western in spirit, it embraces innovative solutions to problems when conventional ideas don't work.

The fundamental breakthrough in public policy analysis comes from interdisciplinary interactions in which scholars from different disciplines calibrate their conceptual tools to meet the needs of the real world problems. This center joins physical scientists with technicians and social scientists to find fresh alternatives to serious policy issues of environment and natural resources, health and welfare and economic development.

A corresponding center, The Center for Global Change at the University of Maryland, started an ambitious program of original research, international conferences and policy analysis. It brings together officials from developing countries to discuss the implications of sea level rise, and organizes workshops for state air control administrators to view emerging clean air technology from the U.S. and European manufacturers. It conducts a symposium on legal and policy implications of global warming sponsored by the University of Maryland Law School.

The Center studies impacts of sea level rise on Chesapeake Bay, including wetland losses. Students at the Law School's Environmental Clinic assist the center in compiling a catalogue of state and local laws aimed to slow global warming including CFC regulations, efficiency standards and reforestation efforts.

GOVERNMENTAL PROBLEM AREAS

Complicating the government and the environment are the following problem areas; human population growth and distribution of wealth, interactions between the ecosystem and socio-economic problems, and the global nature of causes, effects and actions, and policies required to modify the climate.

The population of planet Earth in 1987 passed five billion, and in the next 35 years is expected to reach 8.5 billion. The greatest population increases occur in less developed countries and threaten social and economic stability. Problems emerging from the environment include overlogging caused by expanding populations, and flooding caused by deforestation. Heavy floods from deforestation led Thailand to ban logging and Malaysia to consider this course of action.

The population pressures force subsistence farmers to farm marginal lands—in Brazil the peasants from overpopulated areas destroyed millions of acres of rainforest to subsist on marginal land that rapidly loses its fertility after being cleared and farmed. Because of the debt crisis in Latin America, the bountiful natural resources are used to meet financial obligations to their creditors from other countries. Economic, political and environmental issues are intertwined and cannot be separated.

Another continent adversely affected by overpopulation and the environment is the African continent. Again, smallholder agriculture is ignored, and global economic systems remove much of the local resources. Their soils are as fragile as those found in Latin America rainforests, but drought and desertification threaten their existence and economy. Throughout the world, six million hectares of productive dryland turns into worthless desert through misuse. Poverty becomes a major cause and effect of global environmental problems.

What can be done about overpopulation? Some believe that large numbers of people do little harm to the planet, while others argue that the size of the population should be controlled—these advocates are very vocal. Most of the world's future increase will come from less developed countries, while the developed nation's growth seems to be stabilized. The developed nations need to help developing nations by accepting their exported products. These nations can hinder the less developed regions by imposing prohibitive tariffs on their products and insisting on debt repayment. The eventual hope lies in education against overpopulation which will result in less stress to the environment.

Correlated with the population question is the interaction of the planet's ecosystem and the human socioeconomic system. A prior concern was the rapid change in the ecosystem due to the greenhouse effect, and an inability for current species to cope with these rapid changes, especially belts of plant material. Prior to this century, the natural evolution of species to climate took centuries to evolve. Also, the socioeconomic systems moved slowly in past centuries, but now progresses more rapidly.

However, some facets of this system must evolve more slowly, such as energy technology and the location of populations. Climate change occurs very slowly, and must have long policy intervention and changes in fundamental systems. A facet of this problem concerns the emissions effect which may not be realized until future decades. We are more likely to concentrate on short term effects, than effects which may take years.

This entire book concentrates on global efforts between countries by the use of various international conferences. Some conferences have been effective, while others seems to lack the global cooperation and impetus to make required changes in the environment. Just as the United Nations has an important part to play in maintaining world peace and world health, as it did in eradicating smallpox from all nations, so it remains the hope of the future in formulating environmental changes and in bringing developed and developing countries together in harmonious environmental causes.

Besides cooperation between the governments of various countries, and an international body like the U.N., various technological efforts and scientists play a key role in informing the world of overwhelming problems like the greenhouse effect, ozone depletion, deforestation and acid rain. We must rely on technological efforts, greater fuel efficiency and alternative fuels. Various sections of the book covered fuel efficiencies and scientific efforts. Many countries are concerned about nuclear fission technologies that produce long-lived radioactive wastes for which there are no storage facilities. The resultant low-level radiation products constitute a health problem.

Right now, we are burning fossil fuels that took nature over one million years to produce. Coal, oil, and natural gas supply 88% of global energy and nuclear energy fills in the rest. With the current crisis in the Middle East involving oil-rich countries, we are overly dependent on their oil resources. This dependence creates international crises and potential threats of war. In terms of pollution risk, gas is the cleanest fuel with oil next and coal third. But all create three interrelated atmospheric pollution problems, global warming, urban industrial air pollution and acid rain.

Irving Mintzer of the World Resources Institute remarked, "It is difficult to imagine an issue with more global impacts on human societies and the natural environment than the greenhouse effect. The signal is unclear, but we may already be witnessing examples, if not actual greenhouse effects in Africa ...."

Recommendations in the booklet, "Blueprint for the Environment" include efforts to improve the fuel economy of new cars in the U.S. Currently, the fuel economy of cars is 28 miles per gallon, but the technology makes 60-80 mile per gallon cars possible. We could reduce the country's oil import bill, reduce urban air pollution and slow global warming. Other options include mass transit and passenger railroads.

Alternative fuels are the way to go, and include biofuels, wind energy, geothermal energy, hydroelectric and solar energy. Blueprint recommends that the President of the U.S. needs to increase federal support of research, development and commercialization of energy efficiency and renewable energy sources which have been neglected. We need to remember there are environmental costs associated with fossil fuel use.

Solar energy is an attractive option as the price of this energy continues to drop. Photovoltaic cells produce electricity which converts sunlight to electricity and costs 30 cents per kilowatt hour. Commonly, it is used as a power source for calculators, watches and satellites. Still, photovoltaic cells are more expensive than conventional electricity generation, and we need to increase efficiency and reduce cost. Solar energy can be converted to heat in solar-central-thermal systems. A California company opened a solar-thermal plant where power is produced for less than eight cents per kilowatt-hour.

Wind power may be a new source of energy. A survey of Europe's wind sources published in The Netherlands finds Europe has enough locations to get all the Continent's electricity needs from the wind. Sweden is considering replacing all its nuclear reactors with wind turbines anchored to the sea floor. Visitors to California may be familiar with the wind generators along the highways.

Geothermal energy extracts heat from underground masses of hot rocks, and there are ocean-thermal energy conversion using the temperature difference between the ocean's upper warm water and the cold depths. Recently, Hawaii is developing geothermal power, but conservationists fear the destruction of the U.S.'s last tropical rainforest along with the power production. Iceland has vast resources of underground geothermal production of heat, and heats their buildings and swimming pools with geothermal energy.

One interesting source of energy is the production of energy from methane produced by urban solid waste. There must be a way to separate combustible from non-combustible garbage to make this effort more reliable. Methane or alcohol can enhance and increase the value of traditional fuels. Traditionally, wood has provided energy for much of the world; is renewable since new trees can be grown, and there will always be a supply.

The last alternate energy source is hydroelectric power, but it carries a high capital and environmental cost. As hydroelectric is developed, populations are removed from their homes, dislocated and rainforests are being destroyed. As an alternative, small-scale hydroelectric projects may be useful for developing nations.

12 BLUEPRINT FOR TOMORROW

The Furies were three goddesses of the underworld in Roman mythology, and served as attendants to Proserpina, the queen of the underworld. They acted as sounding boards for complaints against people who committed offenses such as rudeness of the young to the aged, of children to parents and of hosts to guests. Offenders suffered punishment by being driven without rest from city to city and country to country.

The Furies could be likened to our environmental problems, and may be activated against all those people committing environmental ills, today. We could expand the Furies to four instead of three, and could relate them to global warming, ozone depletion, deforestation and overpopulation. Offenders in these areas could be driven from their home countries, and become perpetual wanderers across the Earth.

These Four Furies are everywhere around us. The United States, as an industrialized country and the major contributor to global warming, must take steps to slow this process. Constant stalling is the order of the day, mainly because of the cost of implementation. Current figures estimate that the cost could be as great as the money spent on defense.

Many technologies exist to cut U.S. energy by 50%, but we lack the conservation ethic. James Hansen of NASA remarks, "The danger of crying wolf too soon, which much of the scientific community fears, is that a few cool years may discredit the greenhouse issue... A greater danger is to wait too long..."

The first step in reducing carbon dioxide emissions in the U.S. impacts on increasing the fuel economy of autos and light trucks, and increasing the mileage per gallon to 45 mpg and 35 mpg by the year 2000. Many believe the fuel economy is now sufficient, since in 1989 the cars averaged only 26.5 mpg.

Polluters can be taxed by a boost in the gasoline tax from 9.1 cents a gallon to 50 cents a gallon. The higher taxes could help reduce the Federal budget deficit, and remove our dependence on Gulf Coast oil. The recent 1990 crisis points to our dependence on Arabian oil, and potential consequences which could lead to a Third World War.

Already, steps are under way to add more alternative fuels such as methanol to gasoline. This addition will act in two ways; to conserve the oil-based fuels, and to reduce the load of harmful air pollutants, such as carbon monoxide in the air.

The first conference on the environment, the U.N Conference on the Environment held in Sweden in 1972 helped the U.N. Environmental Program (UNEP) to emerge. Global warming is the foremost issue in the next U.N. meeting commemorating the 20th anniversary of the Swedish conference in 1992. Hopefully, the world's nations will come to some accord on the problem.

The second Fury has more severe consequences if we ignore its potential for damage and consists of CFCs destroying the ozone layer with a possible long term effect. They can last 60-140 years, and will continue to be released from factories, solvents, air conditioners, and refrigerators, long after production is stopped.

As the ozone layer continues to deplete, the EPA reports a high increase in skin cancer among the population of North America. Additional people will suffer from skin cancer in the future, and ozone will deplete phytoplankton productivity which is a blow to the food chain.

Some manufacturers are producing new products to get rid of CFCs, HCFCs and HFCS. New companies with these technologies are Cryodynamics, Inc. of New Jersey producing a refrigerator that cools with helium instead of CFCs, General Cryogenics, Inc. of Texas using liquid nitrogen and liquid carbon dioxide for refrigerated products, and Climatron Corp. of Colorado that makes mobile evaporative cooling air conditioners for cars and buses.

Developing countries look to the U.S. for leadership in environmental issues, and we sometimes aren't a good example. Brazil has received constant criticism on cutting down the rainforest, but claims they have a right to develop their own Amazon region. Brazil notes the U.S. is destroying its Alaskan forest, and cuts down more trees than it replaces. However, President Jose Sarney of Brazil announced a plan in April, 1989 to save the Amazon by creating separate zones for economic development and ecological preservation.

Our own Alaska Tongass National Forest had 50 percent of the most productive forest logged since 1950. Reforestation is not a viable option, since regrowth may take 100 years or more. Norman Myers, an environmental consultant, tells of "hot spots" around the world in danger of destruction. Some threatened rainforest habitats are the Choco of western Colombia, the uplands of western Amazonia, the Atlantic coast of Brazil, Madagascar, eastern Himalayas, the Philippines, Malaysia, northwestern Borneo, Queensland and New Caledonia.

The biological species of the world are threatened by reduction through deforestation, and threatened by climatic warming brought on by the greenhouse effect. Habitat destruction is important to tropical biota, but climatic warming also affects the life in the cold temperate regions and the polar areas. If the climate shifts toward the poles at a rate of 100 kilometers or more a century, many plants and animals might be left behind.

OVERPOPULATION

The last Fury concerns the overpopulation problem, and its effects on global warming and allied ills. Paul and Anne Ehrlich are outstanding spokesmen for the philosophy of small families and respect for the earth—he published *The Population Bomb* in 1968, and now co-authors with his wife, Anne, the sequel, *The Population Explosion*. Since their first book in 1968, the population has escalated from 3.5 billion humans to 5.3 billion people.

A professor at Stanford University, Ehrlich is supplied with bountiful energy and hyper-alertness, exhibited by an ability to do two things at once. He follows the beliefs of British economist, Thomas Malthus, with these comment on the environment. "The absolute limit on any species' success is the carrying capacity of its environment—the maximum number of individuals a habitat can support. Humanity will pay the price for exceeding the carrying capacity of the environment as surely as would a population of Checkerspots."

Ehrlich calls himself a part-time scientist and part-time politician. The Ehrlich's book *The Population Explosion*, not only deals with politics, but is a comprehensive treatise on acid rain, drought, the ozone layer, agriculture, disease and demographics. He mentions that the population bomb has detonated. He warns that 200 million people, most of them children, have died of hunger and disease in the past 22 years, and that 95 million new babies are born each year. He believes children of affluent parents take a greater toll on the environment because they consume more resources and require more material possessions.

Ehrlich's principle opponent is Julian L. Simon, an economics professor at the University of Maryland. Simon, the author of *The Ultimate Resource*, published in 1981, regards population growth as healthy. His theory says gradual population growth doesn't harm the environment, but accelerates its progress.

"I find it difficult to understand how they can see some things only as problems, when I see them as miracles. The fact we can keep five billion people alive now is an incredible accomplishment. We've escaped nature's domination, and all they see is problems. 'Escaped' doesn't mean we've beaten it into submission. It means we've killed the mosquitoes and the smallpox germ."

As a glowing example to his point of view, Simon points to India. In 1968, Ehrlich wrote of the problems of India he doubted it would be self-sufficient in food by 1971. Simon says their population has jumped from 500 million in 1966 to 835 million today, and India manages to feed itself in spite of Ehrlich's predictions.

NEW DIRECTIONS IN ALTERNATE ENERGY

Progress proceeds rapidly with both alternate energy fuels and greater fuel efficiency. Since the U.N. report predicted a rise of 5-10 degrees Fahrenheit by the year 2100, six of seven countries at the Houston Economic Summit in July, 1990 announced major curbs in carbon dioxide emissions over the next 10-15 years.

As various countries make new commitments in energy efficiency and alternate energy sources, new technology emerges, and the cost of removing the greenhouse effect will dive. Germany uses only half as much energy per unit of gross national product as the U.S., and plans to reduce its carbon dioxide emissions by 25% by 2005.

The only hope for cities with severe air pollution lies in alternative fuels according to a EPA emissions specialist. Charles L. Gray says, "Although we have made significant progress in reducing emissions per mile traveled over the last 20 years, the number of cars and the miles they travel has almost doubled in the same time. Unless we continue to reduce the pollution emitted per car, clean air in many cities will continue to be an illusion of the dreamers of the '60s and '70s."

Gray believes the answer lies in alternative fuels, methanol, gasohol and other oxygenated fuels. In his book he says alternative fuels do not produce emissions of benzene, butadiene, gasoline particulates and refueling vapors. "Alternative fuels can provide 90-95% reductions in motor-vehicle-associated cancer cases in a cost-effective manner."

Ken Bossong, the director of Public Citizen's Critical Mass Energy project for the past year comments, "Now, a very large percentage of our automotive fleet is using some mix of alcohol in their fuel. Gasohol is a mixture of 90% gasoline and 10% alcohol fuel. And roughly 10% of the nation's cars are using that as fuel. And alcoholic fuels are a biomass technology. In total, alcohol is providing 1% of the nation's energy supply ...."

The top alternative or renewable energy sources according to Ken Bossong are wood waste, the biggest biomass technology uses wood to produce steam and electricity, hydroelectric power accounting for 45% of the U.S's renewable energy and 11-12% of the nation's electricity, wind power with Hawaii and Connecticut the leaders, geothermal tapping the heat energy of the Earth, mostly located on the West Coast, and solar energy with approximately 1.3 million watts installed in the nation since 1990.

A new alternate fuel present in greatest abundance, and very versatile is hydrogen. It could fuel cars, buses, ships, airplanes and homes, and could be made easily from sunlight and water, garbage, weeds or sewage. It could be produced organically or directly from water with the aid of solar power and upon burning could combine with oxygen from the air and turn into water. Recently, a large display of this fuel was exhibited at the University of Arizona sponsored by NASA, the German government and The Hydrogen Association.

Hailed as the answer to energy problems, solar energy research escalated during the oil scare of the 1970s. Gradually, during the Reagan administration, lower oil prices and administration hostility pushed solar technology out of the marketplace. Now in the U.S., there are thousands of rooftop water heaters, and a new megawatt solar power plant by Luz International in California's Mojave Desert. This plant produces 90% of the world's solar electricity, and sells it to Southern California Edison. Luz remains the largest of 35 companies in the solar thermal field.

Solar box cookers can cook most foods in a reasonable amount of time using the sunlight available in hot climate areas. They are easily built with locally available materials, and can help the wood shortage. Several countries are using the solar cookers, such as India with 50,000 cookers and Bolivia.

Another wide-spread use of solar technology is the photovoltaic cells, generating electricity on exposure to sunlight. Because of research, this cost has dropped 80-85% over the last 10 years and sales are up more than 30% in 1988. A remaining problem is the cost which is four to five times more expensive than fossil fuel plants. Still, solar is used for outdoor home lighting, calculators and power in remote areas.

West Germany, Japan and Italy lead the U.S. in sponsoring research and development of solar power. Siemens, a West Germany electronics giant, recently bought out our premiere solar-cell manufacturer, ARCO Solar. Photovoltaics, although started here, are making more money for Germany than for us.

An alternative energy technology mentioned only briefly in this book is the wind energy industry. The American wind industry started in the 1980s, but tumbled when lucrative federal and state tax credits expired. After a number of bankruptcies, the wind industry has regenerated, and the productivity of wind turbines tripled over the past decade. Northern California's Pacific Gas and Electric Co (PG&E) formed a five year, $20 million research and development partnership with the nation's largest wind turbine manufacturer.

California produces 85% of the world's wind energy with 16,000 turbines concentrated in Tehachapi in Palm Springs and The Altamount Pass area—these generate two billion kilowatt hours of electricity yearly. U.S. Windpower of Livermore, California is the largest turbine manufacturer in the world, and is about to develop an agreement with PG&E to develop a "utility grade" windmill, which will generate more electricity.

U.S. Windpower has property in Spain, Taiwan and Inner Mongolia, and is talking with India, China and 23 other nations. Hawaii's major utility operates wind farms producing 1% of the energy used on Oahu and 15% of the power used on the island of Hawaii.

Besides alternative energy options, improved energy efficiency may be the answer to all our energy woes. Amory Lovins, moustached and with thick eyeglasses recently addressed a meeting of the American Association for the Advancement of Science. Improved energy efficiency using today's technology can solve the global warming problem at a profit of a trillion dollars a year, Amory says.

Lovins, wrote *Soft Energy Paths*, a book compared to *Silent Spring* and translated into eight languages. He outlined a scenario in which the nuclear genie is rebottled, and oil, gas and coal are replaced by hydropower, biomass, solar and other sustainable-energy supplies. Lovin says, "I'm not interested in doing with less, but in doing more with less. We don't have to become vegetarians and ride bicycles to save the Earth."

The trillion dollars saved each year is equal to the global military budget. It's more profitable to save energy than build new power plants. Energy efficiency can be so profitable that it ought to be encouraged even if there were no threat of global warming from atmospheric carbon dioxide.

In the United States, we could save one-fourth of U.S. electricity if we used more efficient lighting, more efficient motors and more efficient appliances and other equipment. Sweden found over a 20-year period, they could expand their economy by 50%, reduce carbon dioxide by one-third and lower utility bills, all through increased energy efficiency.

THE GREEN MOVEMENT

One of the fastest growing environmental movements remains the Greens founded in West Germany ten years ago. Currently, there are organizations in 25 countries; they are the only political party who put the environment first. Germany, as a wealthy superpower when the Greens were formed, suffered from smog, half of their trees were dead from air pollution, and the Rhine River was severely contaminated. The two facets of material wealth and environmental poverty symbolized the loss of natural values to the people of the Greens movement.

The Greens were brought together when the Bonn government announced plans to build a nuclear power plant in 1975 in the town of Wyhl, Baden-Wurttemburg. Immediately, opposition arose from farmers and residents who warned that radiation could contaminate the water and crops, and activists joined the residents who formed the first non-violent environmental protests in West Germany. Nuclear power represented excess consumerism and environmental indifference. Later, other protests occurred at a nuclear plant at Brokdorf in Schleswig, Holstein in 1977, and a nuclear-waste reprocessing plant at Gorleben in Lower Saxony in 1978.

By 1979, The Greens had candidates ready for election to the European Parliament, and made a showing at the election of 3.2% of the vote. Because German law recompenses any party winning more than 0.5% of the vote, they received 3.5 deutsche marks, per vote (since 1983, 5 DM). Their windfall of 4.5 million DM was a huge amount to environmental activists. Next year, their political party became "the Greens", and their platform was four pillared: ecology, social responsibility, grass-roots democracy and non-violence. Later; they won 5.6% of the vote and 27 seats in the German election.

Problems arose in their party because they preached "grassroots" democracy, but practiced central planning with government regulation of everything—because of this many became alienated. The "Greens" left many of the compelling ecological issues, and were soon taken over by Christian Democrats and Social Democrats. By 1989, the Greens appealed to only a few voters, and worried about losing their seats in Parliament in the next election.

What part will the Greens play in the future? Jochen Viehauer, former Greens representative in Parliament, explains they are part of a broad, international trend affecting many in all countries. He says ecological issues will dominate world politics in the next century because there is a major environmental crisis. Industrialized countries can't continue to grow at the expense of Earth and her resources.

Because of the Green movement, West Germany has worked diligently on its environmental ills. They instituted a program to fight acid rain, broke a deadlock during negotiations to protect the ozone layer, and are expected to reduce their carbon dioxide emissions substantially.

NEW ENVIRONMENTAL WOES

Since this last chapter zeros in on new problems and solutions arising since the inception of this book, two of the foremost are the ecological disasters found in Eastern Europe, and problems of nuclear waste disposal. As communism has retreated from the Soviet union, a more open attitude shows in disaster areas where Communist governments emphasized industrialization over the environment.

Imagine smokestacks in our country pouring unfiltered smoke into the air, and pipes dumping untreated wastes into the water. East Germany and Poland lead with acid rain, and other countries of Czechoslavakia, Poland and Hungary have works of art turning into unrecognizable objects. Besides the problem in East Germany, Poland's Sudety Mountains have large areas of forest eaten away by acid rain.

The world's deepest lake, Lake Baikal in the Soviet Union, was a pristine lake with clear water and many species of unique animals. After the government made plans to build a cellulose plant with wastewater released into the lake, many Soviets protested the intrusion and destruction of Lake Baikal. The by-products from the cellulose plants were destroying the lake. Finally, Gorbachev decided to replace the plants with a non-polluting furniture factory.

New environmental reformers are arising in the Eastern Block countries. Some are forming "Green Libraries" to help citizens understand their environment. Other activists mobilize to protest interdependent dams and hydropower plants on the Danube River. Their protests are stopping some of these projects. Sweden helps poorer nations alleviate water pollution in the Baltic Sea by setting up $150,000 to start a network of water quality testing stations, and another $50 million to Swedish-Polish cleanup projects.

Dumps for low-level and high-level radioactive waste continue to be a problem area, since no one wants them located in their back yard. Three disposal sites in the U.S. for low-level waste are located in Nevada, Washington, and South Carolina, but the residents of these states no longer want to be the nation's receptacle for nuclear waste.

California searched for a place to put their low-level radioactive waste since 1980, when Congress advised states they must locate a disposal site. The state selected U.S. Ecology to build a $30 million dump to operate for three decades and then be covered up. The site selected was Ward Valley, 24 miles west of Needles in the Mojave Desert, because of its low annual rainfall and a ground water level of 700 feet below the surface.

At first the residents of Needles agreed to the dump, but after more thought decided to oppose the whole project. Charles Butler, a retired engineer formed PARD—People Against Radiation Dumps, decided to postpone his retirement in favor of beating the nuclear dump. Butler says, "I came out here to retire, but instead I'm spending every hour of the day on the phone. We're going to beat this thing no matter what it takes."

PARD discovered U.S. Ecology had a checkered history, and in the past was sued by the state of Illinois after radioactivity leaked from a closed low-level radioactive dump site. U.S. Ecology says the Illinois environment of heavy rain and shallow ground water table is different, and would not affect the dump site in Needles. Dump opponents worry that trucks may have accidents and spill the low-level radioactive waste.

A NEW CUTTING EDGE TECHNOLOGY

In the chapter on Global Warming Scientists, James Lovelock had a Gaia hypothesis on the effect of phytoplankton on the earth's climate, related to a chemical called dimethyl sulfide. Perhaps their strongest impact on the Earth's climate is their presence in the ocean where they remove one-half of the carbon dioxide we put into the air. Satellite pictures have shown carpets of phytoplankton in the world's oceans, where they have a tremendous effect on the greenhouse effect.

Plant plankton are single living cells of microscopic dimensions— their home is the ocean as far as sunlight can penetrate. Their name is "phyto" from the Greek word for plant and "plankton" for drifting. All the life in the ocean depends on phytoplankton because they convert sun's energy, carbon dioxide and water into potential food for other life species.

Bigelow Laboratories of Ocean Sciences in West Boothbay Harbor, Maine is the location for scientific research on the effects of phytoplankton, and is the home for the discovery of a new organism, *Synechoccus*, a naked chloroplast, a pure photosynthesis machine. This plant may have removed carbon dioxide from the Earth's first primitive atmosphere as increased carbon dioxide caused the air to lack breathable qualities.

Because we are dumping 6 billion tons of carbon dioxide into the atmosphere yearly, and only half shows up in the atmosphere as increased carbon dioxide, we assume that the phytoplankton remove all the excess carbon dioxide.

Theories abound hypothesizing whether increased carbon dioxide could affect the phytoplankton positively, and could help them to pump more gases out of the atmosphere. Conversely, a warming of the North Atlantic could slow the bloom of phytoplankton and accelerate global warming problems. No one knows for sure of the potential effect.

A person who reflects about the state of the planet is former Senator Gaylord Nelson of Wisconsin, who was the founder of the original Earth Day. After a lifetime, he meditates about the imperiled planet, and the politics surrounding the new conservation movement. He mentions his boyhood in Clear Lake, Wisconsin, a town of seven hundred people, "Where I grew up, you lived outdoors. You got an interest in the environment through osmosis."

He says the public's sensitivity to the environment grows yearly. "Fewer and fewer people hold to the old belief that the world is too huge to be damaged by us. But we've only slowed down the rate of degradation. We haven't turned things around... The biggest problem we have now is still the dearth of political leadership ..... The only leader who's been really steadfast on this is the former prime minister of Norway. I'm now thinking about a worldwide demonstration that forces this issue onto the agenda of politicians all over the world."

Senator Nelson's remarks may well reflect the opinions of thousands of environmentally conscious people as the twenty first century looms on the horizon.

BIBLIOGRAPHY

CHAPTER 1

Bagwell, Keith. "Mystery Surrounds Radioactivity Study." Arizona Daily Star, 11 June 1989.

Schneider, Stephen H. "Doing Something About the Weather." "World Monitor," December 1988.

Sierra Club. "1989-90 Conservation Campaign: Global Warming." San Francisco, California: May 1989.

Sierra Club. "Global Warming—Climate in Crisis." San Francisco: July 1989.

Sierra Club. "Sweet Water- Bitter Rain—Toxic Air Pollution in the Great Lakes Basin." Madison, Wisconsin. May 1988.

Ibid. 1989. Update.

CHAPTER 2

Brown, Michael H. The Toxic Cloud. New york, Cambridge, Philadelphia: Perennial Library, Harper and Row, 1987.

Ehrlich, Anne H. and Ehrlich, Paul R. Earth. New York: Franklin Watts, 1987.

Schneider, Stephen H. Global Warming. San Francisco: Sierra Club Books, 1989.

Weir, David. The Bhopal Syndrome. San Francisco: Sierra Club Books, 1987.

CHAPTER 3

Budyko, M.I. The Earth's Climate: Past and Future. International Geophysic Series. New York, London: Academic Press, 1982, Vol.29.

Gribbin, John. The Hole in the Sky. Toronto, New York, London: Bantam Books, 1988.

Regan, James L., Rycroft, Robert W. The Acid Rain Controversy. Pittsburgh: University of Pittsburgh Press, 1988.

CHAPTER 4

"Audubon Activist." Global Warming Issue. January/February 1990.

Beard, Betty. "Valley's Switch to Oxygenated Fuels is On." Arizona Republic. 1 October 1989.

Cubberly, Pamela S. "The Consequences of Global Warming Warming for Biological Diversity." World Wildlife Fund Letter, #5, 1989.

"EPA's Plan for Cooling the Global Greenhouse." "Science," Vol. 243, March 1989, p. 1544.

Fisher, Arthur. "Inside the Greenhouse." "Popular Science," August 1989, p. 63.

Fisher, Arthur. "Playing Dice With the Earth's Climate." "Popular Science," August 1989, p. 51.

Ingram, Helen and Mintzer, Carole. "How Atmospheric Research Changed the Political Climate."

Leatherman, Stephen P., Ph.D. "Impact of Climate-Induced Sea level Rise on Coastal Areas." Presented to United States Senate Committee on Commerce, Science and Transportation. July 13, 1988.

Lemonick, Michael D. "Feeling the Heat." "Time," 2 January 1989, p. 36.

Manabe, Syukuro and Wetherald, Richard T. "CO2 and Hydrology." "Advances in Geophysics." Vol. 28A, 1985.

Maunder, W.J. The Human Impact of Climate Uncertainty. London and New York: Routledge, 1989.

Schneider, Stephen H. "The Changing Climate." "Scientific American," September 1989, p. 70.

Schneider, Stephen H. "The Greenhouse Effect: Science and Publicity." "Science," February 10, 1989, Vol. 243, p. 771.

"Science News." "Rising Seas May Herald Global Warming." June 10 1989, Vol 135:367.

Stavins, Robert N. "Harnessing Market Forces to Protect Environment." January/February 1989, Vol.31, #1, p. 5.

Union of Concerned Scientists. "Renewable Energy and Global Warming." Cambridge, Massachusetts, September 1989.

Udall, James R. "Turning Down the Heat." "Sierra," July/August 1989, p.26.

United Nations Environment Programme. "The Greenhouse Gases." Nairobi, Kenya: UNEP, 1987.

Usher, Peter. "World Conference On the Changing Atmosphere: Implications for Global Security." "Environment", January/February 1989, vol.31, #1, P.25.

CHAPTER 5

Arizona Daily Star. "Group Urges Phase-out of Common Chemicals." 19 June 1990.

Caldwell, Martyn M., Teramura, Alan H., and Tevina, Manfred "The Changing Solar Ultraviolet Climate and the Ecological Consequences for Higher Plants." "Trends in Ecology and Evolution." December 1989, Vol. 4, #12.

Castro, Janice. "One Big Mac, Hold the Box." "Time," 25 June 1990 .

Dennison, Meg. "Vermont to Ban Freon in Car Coolers." "Arizona Republic", 10 May 1989.

Dejen, Jim. "Momentum Builds To Ban CFCs." "Philadelphia Inquirer," 12 March 1989.

Dumanoski, Dianne. "Plan Adopted to Curb Chemicals harmful to Ozone." "Boston Globe," 28 January 1989.

Environmental Defense Fund. "Protecting the Ozone Layer: What You Can Do." New York City: 1988.

Fisher, David E. Fire and Ice. New York: Harper and Row, 1990.

Galloway, Barbara. "Environmentalists Critical of EPA's Ozone Regulations." "The Beacon Journal," Akron, Ohio. August 8 1988.

Gribben, John. The Hole in the Sky. Toronto, New York, London: Bantam Books. 1988.

Lemonick, Michael D. "Letting the Earth Breathe Easier." "Time," 9 July 1990.

Lyman, Francesca, et al. The Greenhouse Trap. World Resources Institute. Boston:Beacon Press, 1990.

Merikaarta, Kaarina. "Ozone Away." "Environmental Action." March/April 1989, p.14.

Miller, Alan. Executive Director of the Center for Global Change, University of Maryland. Letter of April 1990.

"Not Man Apart." "Ozone Campaign News." Friends of the Earth. February-May, 1989, p. 8.

Rowland, F. Sherwood. Testimony Before the Senate Subcommittee on Environmental Pollution. June 1986.

"Science News." "A Promising Alternative to CFC-113." 3 February 1990.

"Science News." "Antarctic Ozone Hole Unexpectedly Severe." 14 October 1989, 136:246.

"Science News." "Ozone Hole Threatens Polar Plankton." 28 October 1989, Vol. 136:284.

"Science News." "Winter zone Gap Detected Over Arctic." July 22 1989, Vol. 136:54.

Shell, Ellen Ruppel. "Solo Flight Into the Ozone Hole Reveals Its Cause." "Smithsonian," February 1988, p. 142.

Weiner, Jonathan. The Next One Hundred Years. Shaping the Fate of Our Living Earth. New York, London, Toronto: Bantam Books, 1990.

CHAPTER 6

Bogart, Herman L. Interview. 26 July 1990.

Bogart, Herman L. Letter written March 1990. Allendale, NJ.

"Desert Voices." Las Vegas, Spring 1990, #8.

"Desert Voices." Las Vegas, Summer 1990 #9.

Ecker, Martin. D., M.D. and Branesco, Norton J., Radiation All You Need to Know to Stop Worrying ...Or To Start. New York: Vintage Books, 1981.

Fradkin, Philip L. Fallout—An American Nuclear Tragedy. Tucson: University of Arizona Press, 1989.

Gale, Dr. Robert Peter, Hauser, Thomas. Final Warning—The Legacy of Chernobyl. New York : Warner Books, 1988.

Gould, J.M., Sternglass, Ernest J. "Low-Level Radiation and Mortality, " "Chemtech, " January 1989, Vol. 19.

Holt, Maria, Interview. 11 July, 1990.

Holt, Maria. Personal Correspondence about Maine Yankee Plant.

Holt, Maria, and King, Elisabeth. Monitoring Maine Yankee. Report of the Citizen's Monitoring Network 1979-1988. Bath, Maine, August 1988.

Johnsrud, Judith H., Ph.D. "On the Trail of Childhood Cancers." "East-West," November 1987. P. 43.

Driesberg, Joseph. "Shutdown Strategies—Citizen Efforts to Close Nuclear Power Plants" public Citizen Critical Mass Energy Project. Washington, D.C.

May, John. The Greenpeace Book of the Nuclear Age— The Hidden History, The Human Cost. New York, Canada: Pantheon Books, 1989.

Null, Gary. Clearer, Cleaner, Safer, Greener. New York: Omni Books Villard Books, 1990.

Sternglass, Ernest J., Ph.D. Secret Fallout, Low Level Radiation from Hiroshima to Three Mile Island. New York, Montreal, London: McGraw Hill Book Co. 1981

Wasserman, Harvey and Solomon, Norman. Killing our Own. The Disaster of America's Experience with Atomic Radiation. New York: Delacorte Press, 1982.

Weart, Spencer R. Nuclear Fear—A History of Images. Cambridge: Harvard University Press, 1988.

Wiley, John K. "Radiation Leak Report Brings Mixed Feelings to Hanford Neighbors." "Arizona Daily Star," 13 July 1990.

"Women's International Coalition to Stop Making Radioactive Waste," Hewitt, New Jersey, June 1990.

Gofman, John W., Ph.D. Letter, Committee for Nuclear Responsibility, Inc. San Francisco, April 1988.

CHAPTER 7

AQMD. "Advisor." Public Advisor's Office. Office of Public Affairs for the Small Business Assistance Section, El Monte, California. October 1989.

AQMD. "Commuter Briefs." El Monte, California. April/May/June 1989.

Carr, Donald E. The Breath of Life. New York: W.W. Norton and Co., Inc., 1965.

Bibliography

Done, Marlu. "Blueprint; for Clear Skies." "Sierra," July/August 1989, p. 16.

Edelstein, Michael R. <u>Contaminated Communities— The Social and Psychological Impacts of Residential Toxic Exposure</u>. Boulder and London: Westview Press 1988.

Franklin, Ben A. "In the Shadow of the Valley." "Sierra," May/June 1986.

French, Hilary F. <u>Clearing the Air: A Global Agenda</u>. Worldwatch Institute, 1990.

Getlin, Josh. "Mr. Clean's Air Act." "Sierra," November/December, 1989, P. 77.

Hall, Dee J. "Boomtown, Power Plants Are At Odds." "Arizona Republic," 22 October 1989.

Levy, Paula. "Where Does it Hurt?" AQMD. El Monte, California, March, 1989.

Nintzel, Jim and Pacenti, John. "Toxic Overload—How the Arizona DEQ Dug Itself Into a Great Big Hole in Mobile." "The Tucson Weekly," 11 July-17 July, 1990.

"Not Man Apart." "Deadly Fertilizer to Keep its Friendly Label." Friends of the Earth, June-September, 1989, p. 9.

Ostmen, Robert Jr. <u>Acid Rain—A Plague Upon the Waters</u>. Dillon Press, Inc., 1982.

Smith, Dr. George. Personal correspondence. 6 March, 1990.

Stewart, Robert W. "Uncertain Cost Clouds Clean Air Debate." "Los Angeles Times," 20 September 1990.

Tinianow, Jerry. Interview. January, 1990.

Wilson, Larry Dr. "The Arizona Toxic Waste Controversy." "Arizona Networking News." Scottsdale, Arizona, Summer, 1990, Vol. 9, No. 2.

Ziehe, Helmut. Interview on August 1990.

CHAPTER 8

Associated Press. "Scope of Acid Rain Study Extends from Lakes to Cars." "Arizona Daily Star," February 21, 1990.

Audette, Rose Marie I. "Acid Rain is Killing More Than Lakes and Trees." "Environmental Action," May/June 1987, p. 10.

Baines, John. Conserving Our World—Acid Rain. Austin: Steck-Vaughn Library, 1989.

Canadian Embassy. "Canada-United States Acid Rain." July 1989.

Environment Canada. "The Non-Ferrous Smelting Industry—Controls to Reduce Acid Rain." 1989.

Hebert, H. Josef. "Senate OKs Bill to Beef Up Clean Air Act." "Arizona Daily Star." 4 April 1990.

International Conference of Ministers on Acid Rain. Environment Canada. March 1984. Ottawa, Canada.

Baker, Joan P., et al. "Effects of Acidic Deposition on Canadian Ecosystems: A Review." Acid Deposition Utility Air Regulatory Group. 22 October 1987.

Munson, Richard. "Will Coal Research Clean the Air?" "Sierra," July/August 1986.

Ostmann, Robert Jr. Acid Rain—A Plague Upon the Waters. Minneapolis: Dillon Press, inc. 1982.

Pringle, Laurence. Rain of Troubles— The Science and Politics of Acid Rain. London: Macmillan Publishing Co., 1988.

Shaw, Robert W. "Air Pollution by Particles." "Scientific American," August 1987, p.96.

"U.S.A. Today." Editorial. "Fight Acid Rain; Make Polluters Pay," 6 March 1990.

Wilford, John Noble. "Refinery, Tourist Bus Exhaust Drench Maya Temples with Ruinous Acid Rain." "Arizona Daily Star," 9 August 1989.

CHAPTER 9

"E" Magazine. "Speaking for the Trees." March/April 1990. P. 42.

Edlin, Herbert, Mimmo, Maurice, et al. The Illustrated Encyclopedia of Trees.

Environmental Defense Fund. "Changing Environment." 1989-90 Annual Report, 1989.

Environmental Defense Fund Letter to Mr. E. Patrick Coady. Executive Director World Bank, 9 January 1990.

Goldstein, Linda. "Redwood Ruckus—Loggers on One Side, Earth First! On the Other." "Arizona Daily Star." 26 May 1990.

Graedel, Thomas E. and Crutzen, Paul J. "The Changing Atmosphere," "Scientific American," September 1989, vol 261, #3, p. 108.

Lester, Robert. "Tropical Deforestation: Chopping Away at Climate Stability." "Audubon Activist," September/October 1989, p. 10.

Linden, Eugene. "Playing with Fire." "Time," 18 September, 1989.

Native Forest Council. "Forest Voices." Eugene, Volume 1, no. 1 1989.

Page, Jake and the Editors of Time-Life Books. Planet Earth Forest. Alexandria: Time-Life Books 1983.

Page, Kara D. "Big Projects, Big Problems." "E" Magazine, May/June 1990, p.30.

Perkins, Nancy. "How One Rainforest Was Saved." "Greenpeace," May/June 1989, p. 18.

Pokhriyal, T.C. and Rao, B.K. Subba. "Role of Forests in Mitigating Air Pollution." "Indian Forester," July 1986, Vol. 112, No. 7.

Rich, Bruce M. Sept. 26 1989. Statement of Bruce M. Rich. Concerning Environmental Performance of the World Bank. Before Subcommittee of International Economic Policy and Trade Subcommittee Human Rights and International Organization. U.S. House of Representatives.

Rich, Bruce. "Conservation Woes at the World Bank." "The Nation," 23 January 1989.

Schwartzman, Dr. Stephan. "Letter for Environmental Defense Fund." 1989.

Richards, John F. and Tucker, Richard P. Editors. World Deforestation in the Twentieth Century. Durham and London: Duke University Press 1988.

Shabecoff, Philip. "Group Warns of Severe Decline in Tropical Forests." "Arizona Daily Star," 8 May 1990.

Skow, John "In Washington; Lighthawk Counts the Clear-Cuts" "Time." 29 August 1988, Vol.132 #9.

Sierra Club. "Bankrolling Disasters. International Development Banks and the Global Environment." San Francisco 1986.

Stewartt, Michael M. and Wotkyns, Steele. "No Hiding From High-Flying Conservation Group." "The Post" Blissfield, MI March 1990.

Trees for Tucson/Global Releaf. Factsheet. November 1989.

Tucson Clean and Beautiful. 1988 Annual Review.

Udall, Lori. Statement on Environmental and Social Impacts of the World Bank Financed Sardar Sarovar Dam in India Before the Subcommittee on Natural Resources, Agricultural Research and Environment Commission. Environmental Defense Fund 24 October 1989.

Wilson, Edward O. "Threats to Biodiversity." "Scientific American," September 1989, Vol. 261, #3 P. 58.

Wotkyns, Steele. Director of Development of Lighthawk. Interview on 20 June 1990.

Wood, Daniel B. "Aerial Crusader." "Christian Science Monitor," July 1988.

CHAPTER 10

Hayes, Denis. "Twenty Years Later—Two Wishes for Earth Day 1990," "E" Magazine, March/April 1990, p. 13.

Earth Day 1990. "Earthline." Stanford, California, Issue. 1990.

Nelson, Gaylord. "Earth Day, Then and Now: Reflections by its Founder," "E" Magazine, March/April 1990, P. 58.

Sierra Club. "The Sierra Club: A Guide." San Francisco, CA: Sierra 1989.

CHAPTER 11

Environment Canada. "Stopping Acid Rain—Steps Toward a Bilateral Solution," April 1990.

International Conference of Ministers on Acid Rain. March 1984, Ottawa, Canada.

Keyfitz, Nathan. "The Growing Human Population." "Scientific American," September 1989, Vol. 261, #3, P. 118.

Gibbons, John H., Blair, Peter D., Gwin, Holly L. "Strategies for Energy Use," "Scientific American," September 1989, Vol. 261, #3.

La Bastille, Anne. "The International Acid Test," "Sierra," May/June 1986.

Maize, Kennedy P. Editor <u>Blueprint for the Environment</u>. Washington, DC, November 1988.

Miller, Alan S. "Three Reports on Japan and the Global Environment," "Environment," July/August 1989, Vol.31, #6.

"North American News.". U.N. Environment Programme. August 1990, Vol. 5, #4.

Skolnikoff, Eugene B. "The Policy Gridlock On Global Warming," "Foreign Policy," Summer 1990, #79.

South Coast Air Quality Management District. "Rules—Why We Need Air Quality Rules," El Monte, California.

United Nations Environment Program. "UNEP Profile," United Nations, New York, NY.

World Commission on Environment and Development. <u>Our Common Future</u>. Oxford, New York, Toronto: Oxford University Press. 1987.

CHAPTER 12

Berreby, David. "The Numbers Game.""Discover." April 1990, Vol. 11, No. 4, p. 42.

Chase, Alston. "It's Not Easy Being Green." "America Way." 15 April 1990, p. 57.

Christrup, Judy. "The World Can't Wait for DuPont." "Greenpeace," July/August 1990, p. 18.

Dillon, Time. "USA Can Wean Itself From Oil Dependency." "USA Today," 15 August 1990.

Doniger, David and Miller, Alan. "Cooling the Greenhouse." "Arizona Daily Star," 12 August 1990.

Kunzig, Robert. "Invisible Garden." "Discover," April 1990, Vol. 11, #4, P. 66.

"Los Angeles Times." "Needles: Nuclear Dump Loses Appeal," 20 September 1990.

Martin, Jim. "Has the Solar Age Finally Arrived." "Utne Reader," September/October 1990.

Warren, Jenifer. "Wind Shift—New Life Breathed Into. Once Failing Energy Industry." "Los Angeles Times." 24 December 1989.

Waters, Tom. "Ecoglastnost." "Discover," April 1990, Vol. 11. #4, P. 51.

Wilkins, Roger. "Honor the Earth." "Mother Jones," April/May 1990, p 12.

Wilson, Edward O. "Threats to Biodiversity." "Scientific American," September 1989, p. 108.

ENVIRONMENTAL ORGANIZATIONS

Audubon Activist
950 Third Ave.
New York, NY 10022

Center for Global Change
University of Maryland
College Park, MD

Citizen Clearinghouse for Hazardous Waste
Box 926
Arlington, VA 22216

Committee for Nuclear Responsibility, Inc.
Correspondence P.O.B. 11207
San Francisco, CA 94101

Critical Mass Energy Project
215 Pennsylvania Ave.SE
Washington, DC 20003

Earth First!
Box 5871
Tucson, AZ 85703

Earth Line—Earth Day 1990
P.O. Box AA
Stanford, CA 94309

Citizen's Energy Council
77 Homewood Ave.
Allendale, NJ 07401

Environmental Defense Fund
257 Park Ave. South
New York, NY 10010

Friends of the Earth
218 D Street SE
Washington, DC 20003

Global ReLeaf
P.O. Box 2000
Washington, DC 20013

Greenpeace
1436 U St. NW
Washington, DC 20009

Izaak Walton League of America
140 Wilson Blvd, Level B
Arlington, VA 22209

Lighthawk
P.O. Box 8163
Santa Fe, NM 87504-8163

Native Forest Council
P.O. Box 2171
Eugene, Oregon 97402

National Wildlife Federation
1400 16th St. NW
Washington, DC 20036

Nuclear Information and Resource Service
1424 16th St NW. Suite 601
Washington, DC 20036

Udall Center for Studies in public policy
1031 North Mountain
Tucson, AZ 85721

Union of Concerned Scientists
26 Church St.
Cambridge, Mass. 02238

Wilderness Society
1400 Eye St. NW
Washington, DC 20005

United Nations Environment Programme
DC2-0803 United Nations
New York, NY 10017

World Resources Institute
1709 New York Ave. NW
Washington, DC 20006

Women's International Coalition to Stop Making Radioactive Waste
1936 Greenwood Lake Turnpike
Hewitt, NJ 07421

World Wildlife Fund
1250 24th St. NW
Washington, DC 20037

—OKAY TO COPY COUPON—

OTHER FINE BOOKS FROM R&E ! ! !

BREATH-TAKING: Stopping the Plunder of Our Planet's Air by Edna A. Zeavin. We are running out of our most precious resource—the very air that we breathe. Every day, hundreds of square miles of rain forest are destroyed and thousands of tons of chemical pollutants are dumped into an already fragile atmosphere.

This book will give you the complete picture of an impending disaster. It describes exactly how this situation was created, what scientists have learned about the problem and what steps we must take *now*—before it's too late.

$12.95
Soft Cover

ISBN 0-88247-925-3
Order #925-3

UNIVERSAL KINSHIP: The Bond Between ALL Living Things by The Latham Foundation. Here is a special book that you will want to read over and over again. This collection of inspiring and comforting articles underscores the deep bonds that unite all living things, and shows us how we all can find hope, and greater meaning as we learn to work in harmony with the earth.

Much of the *Universal Kinship* that is explored in this volume centers on the bonds between us and animals. Within the pages of this book, you will learn how pets are now attending school, not as students, but as teachers and therapists. You will discover how infants, the elderly and the sick can benefit from the presence of animals. You will also gain new insights in handling grief after the loss of a loved one. And, you will find critical new information for healing the bond with the planet we live on. This book is a must for everyone. Buy one for yourself, and for someone you love.

$22.95
Cloth Bound

ISBN 0-88247-918-0
Order #918-0

$11.95
Trade Paper

ISBN 0-88247-917-2
Order #917-2

WHAT WORKS: 5 Steps to Personal Power by William A. Courtney. Life is simple—if you know *What Works* and what doesn't. This power packed action guide is a handbook for creating your dreams. Based on time tested universal principles, this book will guide you through the five steps of personal power. Once you master these simple principles, you will be able to create anything you want, from better health, to financial success, to deeper, more loving relationships.

The principles in this book work. William Courtney used them to change his life from one of loneliness and frustration, to one of happiness and fulfillment. Now he is sharing them with you.

$7.95
Trade Paper

ISBN 0-88247-910-5
Order #910-5

THE TAO OF GOLF by Leland T. Lewis. *Why do the greatest golfers all seem to play with such effortless grace?* They have mastered the Tao, or inner game of golf. The Tao is the balance of Yin and Yang, masculine and feminine, active and attractive principles that rule the Universe.

Now you can learn the mental secrets that will dramatically improve your skill and your enjoyment of the game of golf. As you master this noble game, you will discover an increased sense of inner peace and harmony and will learn to master every aspect of your life. Whether you are an occasional duffer or a seasoned pro, this book will show you the way.

$9.95
Soft Cover

ISBN 0-88247-923-7
Order #923-7

THE GOAL BOOK: Your Simple Power Guide to Reach any Goal & Get What You Want by James Hall. Would you like to be able to turn your dreams into realities? You can if you have concrete goals. This book is based upon a unique goal achievement technique developed by a high school teacher and career counselor in California's Silicon Valley. "Action Conditioning Technology" (ACT) will help you convert your dreams and wishful fantasies into obtainable goals. With this new achievement technology, you will be able to decide exactly what you want, what steps you need to take and when you will reach your objective.

$6.95   LC 91-50675   ISBN 0-88247-892-3
Trade Paper   6 x 9   Order #892-3

YOUR ORDER

ORDER #	QTY	UNIT PRICE	TOTAL PRICE

Please rush me the following books. I want to save by ordering three books and receive FREE shipping charges. Orders under 3 books please include $2.50 shipping. CA residents add 8.25% tax.

SHIP TO:

(Please Print) Name: _____
Organization: _____
Address: _____
City/State/Zip: _____

PAYMENT METHOD

☐ Enclosed check or money order
☐ MasterCard   Card Expires _____   Signature _____
☐ Visa

R & E Publishers • P.O. Box 2008 • Saratoga, CA 95070 (408) 866-6303 FAX (408) 866-0825

—OKAY TO COPY COUPON—

OTHER FINE BOOKS FROM R&E ! ! !

BREATH-TAKING: Stopping the Plunder of Our Planet's Air by Edna A. Zeavin. We are running out of our most precious resource—the very air that we breathe. Every day, hundreds of square miles of rain forest are destroyed and thousands of tons of chemical pollutants are dumped into an already fragile atmosphere.

This book will give you the complete picture of an impending disaster. It describes exactly how this situation was created, what scientists have learned about the problem and what steps we must take *now*—before it's too late.

$12.95	ISBN 0-88247-925-3
Soft Cover	Order #925-3

UNIVERSAL KINSHIP: The Bond Between ALL Living Things by The Latham Foundation. Here is a special book that you will want to read over and over again. This collection of inspiring and comforting articles underscores the deep bonds that unite all living things, and shows us how we all can find hope, and greater meaning as we learn to work in harmony with the earth.

Much of the *Universal Kinship* that is explored in this volume centers on the bonds between us and animals. Within the pages of this book, you will learn how pets are now attending school, not as students, but as teachers and therapists. You will discover how infants, the elderly and the sick can benefit from the presence of animals. You will also gain new insights in handling grief after the loss of a loved one. And, you will find critical new information for healing the bond with the planet we live on. This book is a must for everyone. Buy one for yourself, and for someone you love.

$22.95	ISBN 0-88247-918-0
Cloth Bound	Order #918-0
$11.95	ISBN 0-88247-917-2
Trade Paper	Order #917-2

WHAT WORKS: 5 Steps to Personal Power by William A. Courtney. Life is simple—if you know *What Works* and what doesn't. This power packed action guide is a handbook for creating your dreams. Based on time tested universal principles, this book will guide you through the five steps of personal power. Once you master these simple principles, you will be able to create anything you want, from better health, to financial success, to deeper, more loving relationships.

The principles in this book work. William Courtney used them to change his life from one of loneliness and frustration, to one of happiness and fulfillment. Now he is sharing them with you.

$7.95	ISBN 0-88247-910-5
Trade Paper	Order #910-5

THE TAO OF GOLF by Leland T. Lewis. *Why do the greatest golfers all seem to play with such effortless grace?* They have mastered the Tao, or inner game of golf. The Tao is the balance of Yin and Yang, masculine and feminine, active and attractive principles that rule the Universe.

Now you can learn the mental secrets that will dramatically improve your skill and your enjoyment of the game of golf. As you master this noble game, you will discover an increased sense of inner peace and harmony and will learn to master every aspect of your life. Whether you are an occasional duffer or a seasoned pro, this book will show you the way.

$9.95   ISBN 0-88247-923-7
Soft Cover   Order #923-7

THE GOAL BOOK: Your Simple Power Guide to Reach any Goal & Get What You Want by James Hall. Would you like to be able to turn your dreams into realities? You can if you have concrete goals. This book is based upon a unique goal achievement technique developed by a high school teacher and career counselor in California's Silicon Valley. "Action Conditioning Technology" (ACT) will help you convert your dreams and wishful fantasies into obtainable goals. With this new achievement technology, you will be able to decide exactly what you want, what steps you need to take and when you will reach your objective.

$6.95   LC 91-50675   ISBN 0-88247-892-3
Trade Paper   6 x 9   Order #892-3

YOUR ORDER

ORDER #	QTY	UNIT PRICE	TOTAL PRICE

Please rush me the following books. I want to save by ordering three books and receive FREE shipping charges. Orders under 3 books please include $2.50 shipping. CA residents add 8.25% tax.

SHIP TO:

(Please Print) Name: _____

Organization: _____

Address: _____

City/State/Zip: _____

PAYMENT METHOD

☐ Enclosed check or money order

☐ MasterCard   Card Expires _____   Signature _____

☐ Visa | | | | | | | | | | | | | | | |

R & E Publishers • P.O. Box 2008 • Saratoga, CA 95070 (408) 866-6303 FAX (408) 866-0825